International Max Planck Re

at the

Hamburg Studies on Maritime Affairs
Volume 14

Edited by

Jürgen Basedow
Peters Ehlers
Hartmut Graßl
Lars Kaleschke
Hans-Joachim Koch
Doris König
Rainer Lagoni
Gerhard Lammel
Ulrich Magnus
Peter Mankowski
Marian Paschke
Thomas Pohlmann
Uwe Schneider
Jürgen Sündermann
Rüdger Wolfrum
Wilfried Zahel

Malte Müller

A Large Spectrum of Free Oscillations of the World Ocean Including the Full Ocean Loading and Self-attraction Effects

 Springer

Malte Müller
Hamburg University
ZMAW
Bundesstraße 53
20146 Hamburg
malte.mueller@zmaw.de

Dissertation am Departement Geowissenschaften der Universität Hamburg

Erstgutachter: Prof. Dr. Wilfried Zahel (Universität Hamburg)
Zweitgutachter: Prof. Dr. Jürgen Sündermann (Universität Hamburg)
Drittgutachter: Prof. Dr. Maik Thomas (Helmholtz Zentrum Potsdam)

ISBN 978-3-540-85575-0   e-ISBN 978-3-540-85576-7

DOI 10.1007/978-3-540-85576-7

Hamburg Studies on Maritime Affairs ISSN 1614-2462

Library of Congress Control Number: 2008939218

Cover design: WMXDesign GmbH, Heidelberg

Printed on acid-free paper

9 8 7 6 5 4 3 2 1

springer.com

To my parents
Gerhard H. and Barbara Müller

Das Wissen ist eine dünne Eisdecke über dem kochenden Abgrund des Glaubens. Es deckt den Glauben zu, ohne dessen Macht zu erreichen: Der Glaube treibt im Unterbewussten unkontrolliert sein Wesen (...) Ein Gleichnis. Es lohnt sich, bei ihm zu bleiben. Das Eis ist ein Aggregatszustand des Wassers, die Frage stellt sich, ob nicht Wissen ein Aggregatszustand des Glaubens ist.
   (Friedrich Dürrenmatt)

Preface

This book depends on a dissertation prepared at the Departement of Geosciences at the University Hamburg. It was accepted by the Departement with the grade *summa cum laude* in 2008.

I would like to thank my academic advisor *Prof. Dr. Wilfried Zahel* for his constant support and for the long and constructive discussions.

Further, I would like to thank *Prof. Dr. Jürgen Sündermann* for introducing me to the International Max Planck Research School.

The International Max Planck Research School for Maritime Affairs and in particular *Prof. Dr. Dr. h.c. Jürgen Basedow* and his co-directors are thanked for giving me the opportunity to perform this study in Hamburg.

I acknowledge the computational support of the DKRZ and NEC, especially the helpful comments of *Klaus Ketelsen* and *Jens-Olaf Beismann*.

Last but not least many thanks to my wife *Jana Sillmann* and my son *Darius* for giving me the time I needed for this study and providing a joyful and loving home.

This work has been funded by the International Max Planck Research School for Maritime Affairs at the University Hamburg.

Hamburg, August 2008   *Malte Müller*

Contents

Abstract

A new set of barotropic free oscillations of the World Ocean is computed with explicit consideration of dissipative terms and the full ocean loading and self-attraction effect (LSA). This set contains free oscillations that did not appear in the spectra of previous studies. Furthermore, the expansion towards longer periods (165 hours) yields new global planetary modes. Altogether 169 free oscillations are computed with periods longer than 7.7 hours. Of these, 71 are gravitational modes, 92 are topographically controlled vorticity modes, and 6 are planetary vorticity modes. The influence of the LSA is investigated for all three kinds of modes with respect to changes in the periods and in the spatial structure of the sea surface elevation and the horizontal mass transports. In particular, for modes in the semi-diurnal and diurnal period range, the parameterization of the LSA is analyzed.

For the free oscillations in the period range from 9 to 40 hours the corresponding adjoint solutions are computed and used to synthesize semidiurnal and diurnal tides of second degree. Since these free oscillations are determined with and without consideration of the full LSA-effect, this study allows for a detailed analysis of the LSA on the dynamics of ocean tides, e.g. an physical explanation is given for the induced phase delay computed by ocean tide models. Further, the synthesis gives a spectral composition of certain well known tidal features and pairs of free oscillations are identified, diminishing their contribution either on a global or local scale.

Further, semidiurnal and diurnal tidal solutions of a tidal model with assimilation of data are integrated in the procedure of synthesizing tides. This approach shifts the expansion coefficients of each free oscillation in the synthesis of various tidal constituents towards more realistic values and a first attempt is made to improve the eigenfrequencies of the free oscillations through linear least square fits.

Chapter 1
Introduction

In barotropic ocean dynamics the secondary effect of ocean loading and self-attraction (LSA) is known to be an essential part. It is often considered in a simplified manner, because the full LSA-term turns the dynamical equations into an integro-differential equation system that makes consideration of the full effect very time consuming in numerical models. However, a recent review of the LSA-effect recommends that 'most serious applications should use the full integral formulation' [41]. This convolution integral is defined through the so called Green's function of loading and self-attraction. The Green's function are given in terms of spherical harmonics, weighted through the degree-dependent loading Love numbers, which are computed by an earth model considering the features of elasticity and the radial density distribution of the earth [9].

In the application of ocean tide models including the full LSA-effect, the first complete solutions were obtained by [8, 16, 1, 56]. These solutions yield the common result that the main structure of the tidal patterns is preserved when including the LSA-effect, but that the computed tide is generally delayed, and in certain areas significant local modifications are found. Recent analysis of the full LSA-effect and its parameterization in a barotropic ocean model forced by atmospheric wind stress, atmospheric pressure, and tidal forces indicate that there are significant differences in the magnitude of the LSA-term, depending on the time scale of the ocean response [48]. The full consideration of the LSA-effect in circulation models has not been realized so far, but a first investigation of a simplified consideration of the LSA is performed by [50].

For understanding the response behavior of oceanic water masses to atmospheric and tidal forces, knowledge of the barotropic free oscillations is substantial and provides a spectral representation of the LSA-effect in barotropic ocean dynamics. These oscillations consist of gravity and vorticity modes, primarily governed by the gravity of the Earth and by the Coriolis force, respectively. They determine the response of the ocean to tidal forces and are furthermore excited by wind stress and atmospheric pressure.

In the last century approaches to determine the free oscillations of different kinds of schematic basins have been developed. However, in a rotating frame of reference these solutions are not available in a closed form. A detailed representation of the free oscillations of a rotating rectangular basin, including experimental studies, is given by [40]. Free oscillations of the rotating Earth fully and hemispherically covered by water, are obtained by [21] and [22], respectively. These solutions are generalized through the consideration of the full LSA-effect by the studies of [58] and [59], respectively.

The first numerical solutions of free oscillations for a frictionless global ocean with realistic topography on a rigid earth were obtained by [14], [37] (hereinafter referred to as PL1981) and [12]. [27] (hereinafter referred to as MI1989) investigated the long period vorticity modes and the effect of bottom topography with the PL1981 model, but restricted the investigation to the Pacific Ocean. So far, analyzes of the LSA-effect on free oscillations were made only in the abovementioned spherical and a hemispherical model [58] and [59] Analytic solutions depicting the influence of solely the self-attraction effect on the frequency of the free oscillations were given by [24] and for the whole LSA-effect by [23]. All showed a negative frequency shift due to the LSA-effect. [59] indicated that key resonances might significantly move, which consequently changes the rate of tidal power.

A set of free oscillations of a barotropic global ocean model with realistic topography and explicit consideration of dissipative terms was computed by [64] (hereinafter referred to as ZAMU2005). There, the LSA-effect was introduced by parameterization and the full LSA-term was considered for only two modes.

With the inverse iteration method of ZAMU2005 it is not practicable to determine a large number of normal modes when allowing for the full LSA-effect. For this reason, the finite difference model of ZAMU2005 is taken as a basis and is now combined with a new iteration method, the Implicitly Restarted Arnoldi Method [20]. This upgraded version of the eigenmodel is a novel and highly efficient approach to calculate a large spectrum of normal modes of the World Ocean with explicit consideration of the full LSA-effect. Hence, all of the improvements that were suggested in the summary of PL1981 are realized in this new model. These are (i) the inclusion of dissipation; (ii) higher resolution of the bathymetry; (iii) inclusion of additional marginal seas (e.g. Bering Sea); (iv) inclusion of the self-attraction and loading effect; and (v) the inclusion of the Arctic Ocean. Additionally, the new iteration method yielded 94 free oscillations (72 topographical vorticity modes, 21 gravitational vorticity modes, and one planetary vorticity mode), which were not captured by the inverse iteration method used by ZAMU2005. Furthermore, the spectrum has been expanded towards longer periods (up to 165h), yielding new global planetary modes with periods between 4-7 days.

The following chapter (Chapter 2) describes the underlying theory of computing the free oscillations of the World Ocean with explicit consideration of frictional terms and the full LSA-effect. Further, the developed ocean model used for computing the free oscillations is described. Chapter 3 gives an detailed analysis of the computed gravitational modes and vorticity modes. Moreover the effect of LSA on these modes is analyzed. These results are partly published in [30]. In the first part of chapter 4 the free oscillations of the period range from 9 to 40 hours and their corresponding adjoint solutions are used to synthesize the most important semidiurnal and diurnal tides of second degree (Section 4.1), this approach is published in [31]. The second part of chapter 4 describes a first attempt combining the computed free oscillations with the results of an ocean model with assimilation of data.

Chapter 2
Theory and Model

2.1 Theory

The barotropic free oscillations of the global ocean are defined through the linearized homogeneous shallow water equations (e.g. [57]).

$$\frac{\partial \mathbf{v}}{\partial t} + \mathbf{f} \times \mathbf{v} + \frac{r'}{D}\mathbf{v} + \mathbf{F} + g\nabla\zeta + \mathbf{L}_{sek} = 0$$

$$\frac{\partial \zeta}{\partial t} + \nabla \cdot (D\mathbf{v}) = 0, \qquad (2.1)$$

where ζ denotes the sea surface elevation with respect to the moving sea bottom, $\mathbf{v} = (u,v)$ the horizontal current velocity vector. The undisturbed ocean depth is D, the vector of Coriolis acceleration $\mathbf{f} = 2\omega\sin\phi\mathbf{z}$, the coefficient of linear bottom friction r' and the gravitational acceleration g. $\mathbf{F}$ denotes the vector defining the second-order eddy viscosity term $(F_\lambda, F_\phi) = (-A_h\Delta u, -A_h\Delta v)$ and (λ, ϕ) a set of geographic longitude and latitude values. $\mathbf{L}_{sek}$ is the vector of the secondary force of the loading and self-attraction (LSA), it is derived in section 2.1.1.
In spherical coordinates this system of equations is written as:

$$\frac{\partial u}{\partial t} - 2\omega\sin\phi \cdot v + \frac{r'}{D} \cdot u - A_h\Delta_H u + \frac{g}{R\cos\phi}\frac{\partial\zeta}{\partial\lambda} + L_{sek,\lambda} = 0 \qquad (2.2)$$

$$\frac{\partial v}{\partial t} + 2\omega\sin\phi \cdot u + \frac{r'}{D} \cdot v - A_h\Delta_H v + \frac{g}{R}\frac{\partial\zeta}{\partial\phi} + L_{sek,\phi} = 0 \qquad (2.3)$$

$$\frac{\partial\zeta}{\partial t} + \frac{1}{R\cos\phi}\left(\frac{\partial(Du)}{\partial\lambda} + \frac{\partial(Dv\cos\phi)}{\partial\phi}\right) = 0 \qquad (2.4)$$

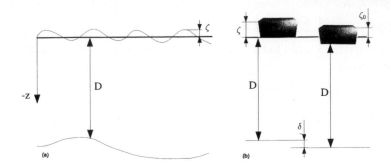

Fig. 2.1 (a) A sketch explaining the used variables z, D, and ζ : the negative z-axis is in downward direction starting from the undisturbed sea-level ($z=0$), the sea surface deformation ζ is in upward direction also starting from ($z=0$). The ocean depth D is the distance between the undisturbed sea-level down to the ocean bottom. (b) **left**: Sea surface deformation ζ without a deformation of the ocean bottom. **right**: The deformation δ of the sea bottom through mass loading. ζ_0 is the geocentric sea level . The distance from the undisturbed sea level to the ocean bottom is now $D-\delta$.

2.1.1 Secondary Forces: The Loading and Self-Attraction Effect

External forces are e.g. tidal forces, atmospheric pressure and wind stress. They have in common that they do not interact in the first order with the ocean dynamics and thus are controlled independently. Against that, the secondary forces have their origin in the dynamics of the water masses and interact with them. In this study the focus is on the secondary forces of loading and self-attraction. The loading-effect results from the deformation of the elastic earth due to the variations of the vertical extension of the water column. The self-attraction is due to the gravitational interaction of the watermasses with themselves. All in all this effect is called the loading and self-attraction effect. This secondary force is derived in terms of the variation of a potential:

Both components of the LSA-effect depend on the sea surface elevation ζ related to the undisturbed sea level and to the actual ocean bottom, i.e. being defined as the variation of the height of the water column (Fig. 2.1a). ζ is written in spherical harmonics:

$$\zeta = \sum_{n=0}^{\infty} \zeta_n = \sum_{n=0}^{\infty} \sum_{s=0}^{n} \overline{P}_{n,s} [C_{n,s} \cos s\lambda + S_{n,s} \sin s\lambda] \tag{2.5}$$

There,

$$\overline{P}_{n,s} = \left(\frac{2(2n+1)}{\delta_s} \frac{(n-s)!}{(n+s)!} \right)^{\frac{1}{2}} P_{n,s}; \qquad \delta_s = 2(s=0), \delta_s = 1(s>0)$$

are the associated Legendre Functions. $C_{n,s}$ and $S_{n,s}$ are time dependent coefficients.

Self-Attraction Effect

Firstly, the elasticity of the earth is neglected, in order to derive solely the expression of the potential of the self-attraction effect. The potential of a spherical layer is given through (Fig. 2.2):

$$\Phi(M) = \gamma \int \int_{\delta K_R} \frac{\rho(M')}{MM'} dS$$

There, δK_R is a spherical layer and $\rho(M')$ is the area density. The distance of the masses M and M' to the center of the sphere is r and R, respectively. Transformation of the reciprocal distance $\overline{MM'}^{-1}$ in spherical harmonics [45] yields:

$$\frac{1}{\overline{MM'}} = (R^2 - 2Rr\cos\alpha + r^2)^{-\frac{1}{2}} = \sum_{n=0}^{\infty} P_n(\cos\alpha) \frac{r^n}{R^{n+1}}, with \quad r < R$$

Using the notation $f(\lambda', \phi') := \rho(M')$ and considering the area elements $dS = R^2 \sin\phi' d\phi' d\lambda'$ yields

$$\Phi(M) = \gamma \sum_{n=0}^{\infty} \frac{r^n}{R^{n+1}} \int \int_{\delta K_R} f(\lambda', \phi') P_n(\cos\alpha) dS.$$

The area density is now written in spherical harmonics $f(\lambda, \phi) = \sum_{n=0}^{\infty} f_n(\lambda, \phi)$ with $f_n(\lambda, \phi) = \frac{2n+1}{4\pi} \int \int_{\delta K_R} f(\lambda', \phi') P_n(\cos\alpha) dS$. Thus, the potential is rewritten in

$$\Phi(M) = 4\pi\gamma \sum_{n=0}^{\infty} \frac{r^n}{R^{n-1}} f_n(\lambda, \phi).$$

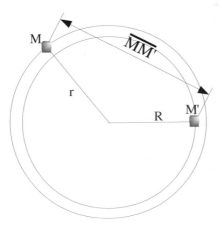

Fig. 2.2 The potential of a spherical layer: The mass M in the potential of the surrounding masses M' ($\Phi(M) = \gamma \int \int_{\delta K_R} \frac{\rho(M')}{MM'} dS$).

Since the radius of the earth R is large compared to the ocean depth, for the radius r holds $r \to R$ and the area density can be approximated through $f(\lambda, \phi) = \rho \zeta(\lambda, \phi)$ with constant distance to the earth's center R. Finally, the potential can be written as

$$\Phi(\phi, \lambda) = g \cdot \sum_{n=0}^{\infty} \frac{3}{2n+1} \frac{\rho_0}{\rho_e} \zeta_n(\phi, \lambda) =: \sum_{n=0}^{\infty} \Phi_n(\phi, \lambda) \qquad (2.6)$$

ρ_0 and ρ_e are the mean densities of the sea water and the solid earth, respectively. This potential describes the so called self-attraction effect. The gradient of this potential is the gravitative force of the watermasses on themselves.

Loading-Effect

Hereafter it is assumed that the earth is fully elastic. Thus, the ocean bottom gets deformed through forces acting on it. In the present study only inner forces can deformate the ocean bottom, external forces are excluded in the homogeneous problem of determining free oscillations. The surface elevation ζ gives rise to two forces; on the one hand the changing weight of the water column, on the other hand the changing gravitational attraction on the ocean bottom.

The spherical harmonic of degree n of the geocentric sea surface elevation is $(\zeta_0)_n$ (compare 2.5), and the deformation of the ocean bottom is described through δ (Fig. 2.1b).

Compared to the undisturbed sea level ($\zeta = 0$), the additional body of water

$$g\rho \zeta_n = g\rho (\zeta_0 - \delta)_n$$

deformates the ocean bottom due to the two abovementioned inner forces. [9] showed that the vertical displacement of the ocean bottom δ_n is proportional to the vertical expansion of the water column. The factor of proportionality is h'_n:

$$\delta_n = h'_n \cdot \frac{3}{2n+1} \frac{\rho_0}{\rho_e} \zeta_n = h'_n \Phi_n / g \qquad (2.7)$$

Since the pressure through the loading overbalances the gravitational attraction, the h'_n are negative for all degrees n. The mass displacement through the deformation of the ocean bottom results additionally in a variation of the potential. This can be described by the factor of proportionality k'_n:

$$V'_n = k'_n \cdot \Phi_n \qquad (2.8)$$

These parameters, h'_n and k'_n, depend on the characteristics of the elasticity of the earth and are called the loading Love-numbers . They describe the interaction of the ocean with the solid earth in terms of spherical harmonics.

The loading Love-numbers used in this study are taken from [10]. They determined the Love-numbers values taking the Preliminary Reference Earth Model (PREM) [5]

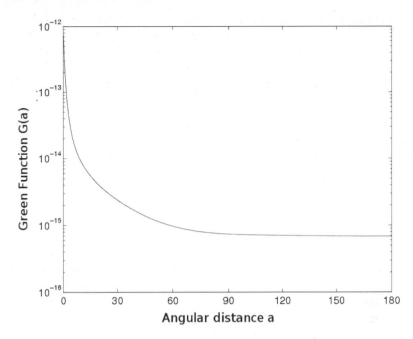

Fig. 2.3 The Green's function depending on the angular distance a between the two points (ϕ, λ) and (ϕ', λ'). It holds for the angular distance: $(\cos(a) = \sin \phi \sin \phi' + \cos \phi \cos \phi' \cos(\lambda - \lambda'))$, (Data from [10]).

as a basis. The Love-numbers of [9], he utilized the Gutenberg-Bullen-Earth-Model, differ distinctly for large degrees n, from the above ones.

2.1.2 The Equations of Motion and the Equation of Continuity

The loading and self-attraction effect is described through the potential (2.8 and 2.6)

$$\Phi^* = V' + \Phi = \sum_{n=0}^{\infty} (1 + k_n') \Phi_n = \sum_{n=0}^{\infty} g(1 + k_n') \alpha_n \zeta_n \tag{2.9}$$

with $\alpha_n = \frac{3}{2n+1} \frac{\rho_0}{\rho_e}$, and the displacement of the ocean bottom δ (2.7).
The secondary force is given by the horizontal gradient ∇_H of this potential (2.9):

$$L_{sek} = \nabla_H \Phi^* =: g \nabla_H \overline{\zeta} \tag{2.10}$$

There, $\overline{\zeta}$ is the equilibrium representation of the secondary potential. The sea level ζ relative to the moving sea bottom is described through the vertical displacement δ of the sea bottom and the geocentric sea level ζ_0:

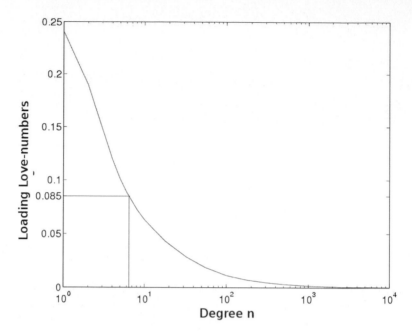

Fig. 2.4 The Love-number combination $(1+k'_n-h'_n)\alpha_n$. The abscissa shows the degree n of the spherical harmonic (Data from [9]). The value $(1+k'_n-h'_n)\alpha_n = 0.085$ is marked, which is often used for the parameterization of the LSA-effect [1].

$$\zeta_0 = \zeta + \delta \tag{2.11}$$

Putting (2.10) and (2.11) in (2.2)- (2.4), results in

$$\frac{\partial u}{\partial t} - 2\omega\sin\phi\cdot v + \frac{r'}{D}u - A_h\Delta u + \frac{g}{R\cos\phi}\frac{\partial\zeta}{\partial\lambda} = \frac{g}{R\cos\phi}\frac{\partial(\overline{\zeta}-\delta)}{\partial\lambda} \tag{2.12}$$

$$\frac{\partial v}{\partial t} + 2\omega\sin\phi\cdot u + \frac{r'}{D}v - A_h\Delta v + \frac{g}{R}\frac{\partial\zeta}{\partial\phi} = \frac{g}{R}\frac{\partial(\overline{\zeta}-\delta)}{\partial\phi} \tag{2.13}$$

To obtain a system of equations with the three state variables u, v and ζ, the spherical harmonics ζ_n must be expressed through ζ. Transformation of (2.5) results in

$$\begin{matrix}C_{n,s}\\S_{n,s}\end{matrix} = \frac{1}{4\pi}\int\int \zeta(t,\lambda',\phi')\overline{P}_{n,s}(\sin\phi')\begin{matrix}\cos(s\lambda')\\\sin(s\lambda')\end{matrix}d\lambda'd\phi'\cos\phi'$$

Finally, by substituting this into (2.12) and (2.13), the system of equations is rewritten as:

$$\frac{\partial u}{\partial t} - 2\omega\sin\phi\cdot v + \frac{r'}{D}u - A_h\Delta u + \frac{g}{R\cos\phi}\frac{\partial\zeta}{\partial\lambda} = \tag{2.14}$$

$$\frac{g}{R\cos\phi}\int\int\zeta(t,\lambda',\phi')\frac{\partial G(\lambda,\phi,\lambda',\phi')}{\partial\lambda}d\lambda'd\phi'\cos\phi'$$

$$\frac{\partial v}{\partial t}+2\omega\sin\phi\cdot u+\frac{r'}{D}v-A_h\Delta v+\frac{g}{R}\frac{\partial\zeta}{\partial\phi}= \qquad (2.15)$$
$$\frac{g}{R}\int\int\zeta(t,\lambda',\phi')\frac{\partial G(\lambda,\phi,\lambda',\phi')}{\partial\phi}d\lambda'd\phi'\cos\phi'$$

$$\frac{\partial\zeta}{\partial t}+\frac{1}{R\cos(\phi)}\left(\frac{\partial(Du)}{\partial\lambda}+\frac{\partial(Dv\cos\phi)}{\partial\phi}\right)=0 \qquad (2.16)$$

This is an integro-differential equation system. The function

$$G(\lambda,\phi,\lambda',\phi'):=$$
$$\frac{1}{4\pi}\sum_{n=0}^{\infty}(1+k_n'-h_n')\alpha_n\sum_{s=0}^{n}\overline{P}_{n,s}(\sin\phi)\overline{P}_{n,s}(\sin\phi')\cos(s(\lambda'-\lambda)) \qquad (2.17)$$

contains the loading Love-numbers. This function is often called the Green's function of loading and self-attraction. An important characteristic of this Green's function is that its dependency on the four variables $(\lambda,\phi,\lambda',\phi')$ can be reduced to that on the angular distance a, given through

$$\cos a = \sin\phi\sin\phi'+\cos\phi\cos\phi'\cos(\lambda-\lambda').$$

This is possible since the Love-numbers, which are forming the basis of the Green's function are only depending on the degree of the spherical harmonics and not on their order. The proof is given through the so called addition-theorem of Legendre-polynomials ([45], page 427):

$$P_n(\cos a)=\sum_{s=0}^{n}\frac{(n-s)!2}{(n+s)!\delta_s}P_{n,s}(\sin\phi')P_{n,s}(\sin\phi)\cos(s(\lambda'-\lambda)) \qquad (2.18)$$

Putting this expression in (2.17), the Green's function is rewritten in the form

$$G(a)=\frac{1}{4\pi}\sum_{n=0}^{\infty}(1+k_n'-h_n')\alpha_n\overline{P}_n(\cos a). \qquad (2.19)$$

The Green's function, determined by the loading Love-numbers of [10], is displayed in Figure (2.3).

2.1.3 Energy Balance

The equations of motion (2.2- 2.4) are transformed into the following energy equation (e.g. [56])[1]:

$$\underbrace{\frac{\partial}{\partial t}[\frac{1}{2}D(u^2+v^2)]}_{Kinetic\ Energy} - \underbrace{\frac{1}{2}\frac{\partial\zeta}{\partial t}(u^2+v^2)}_{Correction\ Term} + \underbrace{r'D(u^2+v^2)^{\frac{3}{2}}}_{Dissip.\ Bottom\ Friction} +$$

$$g\nabla\cdot\underbrace{(Du\zeta_0,Dv\zeta_0)}_{Energy\ Flux} - \underbrace{DA_h((\nabla u)^2+(\nabla v)^2)+DA_h(\nabla\cdot(u\nabla u+v\nabla v))}_{Dissipation\ Through\ Eddy\ Viscosity} +$$

$$\underbrace{\frac{\partial}{\partial t}(\frac{1}{2}g\zeta^2+g\delta\zeta)}_{Potential\ Energy} = \tag{2.20}$$

$$\underbrace{\nabla\cdot(Du\Phi^*,Dv\Phi^*)+\Phi^*\frac{\partial\zeta}{\partial t}}_{Work\ Done\ Through\ LSA-effect} + \underbrace{g\zeta\frac{\partial\delta}{\partial t}}_{Work\ Done\ Through\ Bottom\ Deformation}$$

The terms on the right hand side of this equation, originating from the LSA-effect, are zero in the time-mean energy budget [57].

Now, a particular free oscillation with the frequency $i\sigma = \sigma_1 + i\sigma_2$ and its complex constituents u, v and ζ is considered. The real part of $i\sigma$, i.e. σ_1, determines the damping rate of the mode, with the energy decay time $\frac{1}{2\sigma_1}$ [64]. The eigenperiod $T_2 = \frac{2\pi}{\sigma_2}$ is given through the imaginary part σ_2.

In order to evaluating in (2.20) the potential and kinetic energy contents as well as the energy flux term, the real parts of the constituents u, v and ζ of the complex eigenfunction are used in the form

$$u = |u|e^{-\sigma_1 t}\cos(-\sigma_2 t+\Psi+\phi_u) \tag{2.21}$$

$$v = |v|e^{-\sigma_1 t}\cos(-\sigma_2 t+\Psi+\phi_v) \tag{2.22}$$

$$\zeta = |\zeta|e^{-\sigma_1 t}\cos(-\sigma_2 t+\Psi+\phi_\zeta). \tag{2.23}$$

There, Ψ is an arbitrary phase shift.

The ocean bottom deformation δ is obtained from the sea surface elevation making use of the corresponding Green's function and is likewise represented by the amplitudes and phases

$$\delta = |\delta|e^{-\sigma_1 t}\cos(-\sigma_2 t+\Psi+\phi_\delta).$$

---

[1] In order to obtain energies the equation has to be multiplied by ρ.

The Averaging Method

Considering the time-mean of a product of two periodic functions, e.g. u and ζ (2.21)-(2.23)

$$M_{\zeta u} := \frac{1}{T} \int_0^T dt(\zeta u).$$

Substituting the real functions and integrating over the time results in

$$M_{\zeta u} = \frac{1}{8\pi((\frac{\alpha}{\sigma})^2+1)} \left(1 - e^{\frac{-4\pi\alpha}{\sigma}}\right) [\underbrace{\frac{2\alpha}{\sigma} \cos(\phi_u + \Psi)\cos(\phi_\zeta + \Psi)}_{depending\ on\ \Psi} +$$

$$\frac{\sigma}{\alpha}\cos(\phi_u - \phi_\zeta) - \underbrace{\sin(\phi_u + \phi_\zeta + 2\Psi)}_{depending\ on\ \Psi}]|\zeta||u| \quad (2.24)$$

The eigenfrequency has here the notation $\alpha := \sigma_1$ and $\sigma := \sigma_2$. Obviously, the two marked terms in the equation for $M_{\zeta u}$, depend on the arbitrary phase shift Ψ. Of course, this is due to the damping factor. In case of no dissipation ($\alpha = 0$) these two terms would disappear:

$$M_{\zeta u}^{(without\ dissipation)} = \frac{1}{2}\cos(\phi_u - \phi_\zeta)|\zeta||u|$$

Averaging the resulting $M_{\zeta u}$ (2.24) over the interval $(0, 2\pi)$ with respect to the phase Ψ makes it independent of Ψ:

$$\overline{M}_{\zeta u} = \frac{1}{2\pi} \int_0^{2\pi} d\Psi M_{\zeta u} = B \cdot \cos(\phi_u - \phi_\zeta)|\zeta||u| \quad (2.25)$$

$$B := \frac{1}{8\pi((\frac{\alpha}{\sigma})^2+1)} \left(1 - e^{\frac{-4\pi\alpha}{\sigma}}\right)\left(\frac{\alpha}{\sigma} + \frac{\sigma}{\alpha}\right)$$

Potential and Kinetic Energy

Using (2.20) and (2.25) the time-mean of the potential and kinetic energy surrenders to

$$\overline{E}_p = B \cdot (\frac{1}{2}\rho g|\zeta|^2 + \rho g|\delta||\zeta|\cos(\phi_\delta - \phi_\zeta)) \quad (2.26)$$

$$\overline{E}_k = B \cdot \frac{1}{2}\rho D(|u|^2 + |v|^2), \quad (2.27)$$

The total energy is given through:

$$\overline{E}_t = \overline{E}_p + \overline{E}_k \quad (2.28)$$

Energy Flux

The time-mean of the two components J_u and J_v of the energy flux, is determined through (2.20) and (2.25):

$$\overline{J_u} = \rho g D(\overline{M_{u\zeta}} - \overline{M_{u\delta}}) = B \cdot \rho g D|u|(|\zeta|\cos(\phi_u - \phi_\zeta) - |\delta|\cos(\phi_u - \phi_\delta)) \quad (2.29)$$

$$\overline{J_v} = \rho g D(\overline{M_{v\zeta}} - \overline{M_{v\delta}}) = B \cdot \rho g D|v|(|\zeta|\cos(\phi_v - \phi_\zeta) - |\delta|\cos(\phi_v - \phi_\delta)) \quad (2.30)$$

2.1.4 Parameterization of the LSA - An Analytical Approach

Considering the shallow water equations (2.1) without friction and the LSA-effect, the so called Laplace-equations are given by:

$$\frac{\partial \mathbf{v}_H}{\partial t} + (\mathbf{f} \times \mathbf{v})_H = -g\nabla_H \zeta \quad (2.31)$$

$$\frac{\partial \zeta}{\partial t} + \nabla_H \cdot (D\mathbf{v}_H) = 0 \quad (2.32)$$

$$\mathbf{v}_H \cdot \mathbf{n}|_\Gamma = 0 \quad (2.33)$$

With the Operator $\mathscr{L}_0$

$$\mathscr{L}_0 = \begin{pmatrix} \mathbf{f} \times & g\nabla_H \\ \nabla_H \cdot D & 0 \end{pmatrix} \quad (2.34)$$

and with the vector $\mathbf{w} = \begin{pmatrix} \mathbf{v}_H \\ \zeta \end{pmatrix}$, the above equation system (2.31, 2.32) can be rewritten as:

$$\frac{\partial \mathbf{w}}{\partial t} = -\mathscr{L}_0 \mathbf{w} \quad (2.35)$$

The loading and self-attraction effect is now defined by a perturbation operator $\delta\mathscr{L}$. [23] showed with this perturbation formalism, that the variation of the frequency of a free oscillation through the LSA-effect is

$$\frac{\delta\sigma}{\sigma} = \frac{-\int dS \zeta_{LSA} \zeta^*}{\int dS |\zeta|^2} \cdot \frac{E_p}{E_t} = -\beta \cdot \frac{E_p}{E_t} \quad (2.36)$$

where E_p/E_t is the ratio of the potential energy to the total energy, σ the frequency and ζ^* the conjugate complex sea surface elevation of the free oscillation determined without the LSA-effect. ζ_{LSA} is defined by the LSA-term (compare 2.17):

$$\nabla\zeta_{LSA} = \nabla \int \int_S \zeta(t, \lambda', \phi')G(\lambda, \phi, \lambda', \phi')R^2 \cos(\phi')d\lambda'd\phi'. \quad (2.37)$$

Since ζ and ζ_{LSA} are not exactly in phase the proportional constant β, defined through (2.36), has a complex value. However, the imagenary part is considerably smaller than the real part (the factor is less than 0.001). Therefore β is treated as a real value in the following. As will be shown later, the sign of β is positive and thus the consideration of the LSA-effect results in a decrease of the frequencies of the free oscillations. The relative magnitude of this frequency shift depends on the ratio of potential energy to the total energy and on the factor $\beta = \frac{\int dS \zeta_{LSA} \zeta^*}{\int dS |\zeta|^2}$.

Parameterization of LSA

This β-value is the same, which [1] introduced for parameterizing the LSA-effect in tidal models. There, the LSA-term of equation (2.37) is approximated by: [1] obtained for M_2 constituent $\beta = 0.085$. [33] gave different values for the most important semidiurnal and diurnal tidal constituents (see Table 2.1) and [41] recommends a higher value for the M_2 with $\beta = 0.12$.

The LSA-term is still interactive with ζ when represented in this simple form. Howewer, it is a massive simplification since it does mean for all Love-numbers that $(1 + k'_n - h'_n)\alpha_n = \beta$ (compare Fig. 2.4). Analysis of the local distribution of β shows that there are large differences of β between the open ocean and the coastal region [41, 48]. The values of β are small near land, and are getting large in open ocean areas. Further, [48] introduce a local $\beta_L(\lambda, \phi) = \frac{\zeta_{LSA}(\lambda, \phi)}{\zeta(\lambda, \phi)}$ and discuss its dependency on ocean depth and on latitude for different time scales generated through various forcings (tidal, atmospheric wind, atmospheric pressure) of their barotropic ocean model. For a detailed analysis of the β-values of the gravitational modes see Sections (3.1.1) and (3.1.2), and of the vorticity modes see Section (3.2).

$$\nabla \zeta_{LSA} \approx \beta \cdot \nabla \zeta \tag{2.38}$$

Table 2.1 β-values (2.36) for the most important semidiurnal and diurnal tidal constituents [33].

diurnal		semidiurnal	
K_1	0.121	M_2	0.082
O_1	0.115	S_2	0.083
P_1	0.119	N_2	0.079

2.2 Model

The physical model for the global free oscillations is described through the eigen-value problem

$$-i\mathscr{L}\bar{\mathbf{w}} = \sigma\bar{\mathbf{w}} \tag{2.39}$$

$$\mathbf{v}_H \cdot \mathbf{n}|_r = 0 \tag{2.40}$$

here, the periodic function $\mathbf{w}(\lambda,\phi) = \begin{pmatrix} \mathbf{v}_H \\ \zeta \end{pmatrix} = \begin{pmatrix} \bar{\mathbf{v}}_H \\ \bar{\zeta} \end{pmatrix} \cdot \exp(-i\sigma t)$, with the complex valued frequency $i\sigma = \sigma_1 + i\sigma_2$ is introduced. The operator $\mathscr{L}$, derived from the system of equations (2.14- 2.16), is

$$\mathscr{L} = \begin{pmatrix} \mathbf{f}\times + \frac{r'}{D} - A_h\Delta_H & g\nabla_H - g\nabla_H\mathscr{I} \\ \nabla_H \cdot D & 0 \end{pmatrix}, \tag{2.41}$$

where $\mathscr{I}$ is defined through $\mathscr{I}\zeta = \int G(\lambda,\phi,\lambda',\phi')\zeta(\lambda',\phi')d\lambda'd\phi'\cos\phi'$.

Properly replacing the derivatives by finite differences and the integral by a finite expression [61] makes (2.39) turn into a system of algebraic equations

$$(A - \lambda I)\mathbf{x} = 0 \tag{2.42}$$

where $\lambda = i\sigma = \sigma_1 + i\sigma_2$ represents the eigenvalue of the Matrix A with the corresponding eigenvector $\bar{\mathbf{x}} = \mathbf{x}e^{-i\sigma t}$ depending on space as well as on time, i.e. $\bar{\mathbf{x}} = \bar{\mathbf{x}}(t,\lambda,\phi)$.

The system of equations (2.42) has, in the present case of a spatial resolution of one degree, approximately 120,000 unknowns. Since LSA is taken fully into account the entries of the matrix A are generally nonzero. However, since the Green's function depends only on the angular distance a, symmetries in the arrangement of the entries can be utilized to reduce the working memory of the model to less than 1GB (compare equation 2.19). Taking advantage of this memory reduction, three single free oscillations were computed with a special modification of the Wielandt Method Wielandt Method [64] and four with the standard Wielandt method [29]. In the first case the model was time optimized with respect to the method itself, whereas in the latter one it was distributed with OpenMP on 8 cpus and optimized for the HLRE[2]. Both approaches make use of the Wielandt Method (or inverse iteration), as described in [13] and as originally having been developed by [55], see also [17]. Starting from a first guess eigenvalue σ_0, the method yields the free oscillation with the eigenvalue λ closest to σ_0. The advantage of this method is that single free oscillations are determined with comparable low computational costs due to the possibility of the above mentioned memory reduction. The main disadvantage is the time consuming procedure when allowing for the full LSA-effect and that not all free oscillations are captured by this method.

---

[2] HLRE - High Perfomance Computing Centre for Earth System Research, Hamburg.

2.2.1 The Implicitly Restarted Arnoldi Method

In the present study the Implicitly Restarted Arnoldi Method is used for solving the eigenvalue problem (2.39). It is provided by the software package ARPACK [20]. The original Arnoldi Method [2] is an orthogonal projection method, belonging to the class of Krylov subspace methods. In case of a symmetric Matrix A, it reduces to the Lanczos Method [19]. Below, only a short summary of the Arnoldi method is given. A more comprehensive treatment of the subjects of Krylov subspaces, Arnoldi factorization, and Arnoldi method can be found in [43].

The k-th Krylov subspace associated with the matrix A and the vector $\mathbf{v}$ is defined through

$$\mathcal{K}_k(A, \mathbf{v}) = span\{\mathbf{v}, A\mathbf{v}, A^2\mathbf{v}, ..., A^{k-1}\mathbf{v}\}. \tag{2.43}$$

Obviously, it is defined through the sequence of vectors produced by the power method (e.g. [13]). This method utilizes the fact that with k increasing the vector $A^k\mathbf{v}$ converges to the eigenvector with the largest eigenvalue. Like all Krylov subspace methods, the Arnoldi method takes advantage of the structure of the vectors produced by the power method, and information is extracted to enhance convergence to additional eigenvectors. For this purpose, the Arnoldi method determines an orthonormal basis $span\{\mathbf{u}_1, \mathbf{u}_2, ...\mathbf{u}_k\}$ for $\mathcal{K}_k(A, \mathbf{v})$. This basis is defined through the relation

$$AU_k = U_k H_k + \mathbf{f}_k \mathbf{e}_k^T \tag{2.44}$$

where $A \in \mathbf{C}^{n \times n}$, the matrix $U_k = (\mathbf{u}_1, \mathbf{u}_2, ...\mathbf{u}_k) \in \mathbf{C}^{n \times k}$ (has orthogonal columns), $U_k^H \mathbf{f}_k = 0$, $\mathbf{e}_k \in \mathbf{C}^k$ and $H_k \in \mathbf{C}^{k \times k}$ is upper Hessenberg with non-negative subdiagonal elements. This is called a k-step Arnoldi factorization and its algorithm is shown in Fig. 2.5. Alternatively, the factorization (2.44) can be written as

$$AU_k = (U_k, \mathbf{u}_{k+1}) \begin{pmatrix} H_k \\ \beta_k \mathbf{e}_k^T \end{pmatrix}, \tag{2.45}$$

where $\beta_k = \|\mathbf{f}_k\|$ and $\mathbf{u}_{k+1} = \frac{1}{\beta_k}\mathbf{f}_k$. If $H_k\mathbf{s} = \mathbf{s}\theta$ then the vector $\mathbf{x} = U_k\mathbf{s}$ satisfies

$$\|A\mathbf{x} - \mathbf{x}\theta\| = \|(AU_k - U_k H_k)\mathbf{s}\| = |\beta_k \mathbf{e}_k^T \mathbf{s}|. \tag{2.46}$$

The so called Ritz-pair $(\mathbf{x}, \theta)$ is an approximate eigenpair of A, with the Ritz-estimate as the residual $r(\mathbf{x}) = |\beta_k \mathbf{e}_k^T \mathbf{s}|$ (assuming $\|\mathbf{x}\| = 1$).

Unfortunately, the Arnoldi Method has large storage and computational requirements. Large memory is used to store all the basis vectors u_k, if the number of iteration steps k is getting large before the eigenvalues and eigenvectors of interest are well approximated through the Ritz-pairs. Additionally, the computational cost of solving the Hessenberg eigenvalue subproblem rises with $\mathcal{O}(k^3)$. To overcome these difficulties, methods have been developed to implicitly restart the method [47]. This efficient way to reduce the storage and computational requirements makes the Arnoldi Method suitable for large scale problems. Further, implicit restarting pro-

Input $(A, \mathbf{v})$

Put $\mathbf{u}_1 \mathbf{v}/\|\mathbf{v}\|$; $\mathbf{w} = A\mathbf{u}_1$; $\alpha_1 = \mathbf{u}_1^H \mathbf{w}$;

Put $\mathbf{f}_1 \leftarrow \mathbf{w} - \mathbf{u}_1 \alpha_1$; $U_1 \leftarrow (\mathbf{u}_1)$; $H_1 \leftarrow (\alpha_1)$

For $j = 1, 2, 3, \ldots k - 1$

(1) $\beta_j = \|\mathbf{f}_j\|$; $\mathbf{u}_{j+1} \leftarrow \mathbf{f}_j/\beta_j$;

(2) $U_{j+1} \leftarrow (U_j, \mathbf{u}_{j+1})$; $\widehat{H}_j \leftarrow \begin{pmatrix} H_j \\ \beta_j \mathbf{e}_j^T \end{pmatrix}$;

(3) $\mathbf{w} \leftarrow A\mathbf{u}_{j+1}$;

(4) $\mathbf{h} \leftarrow U_{j+1}^H \mathbf{w}$; $\mathbf{f}_{j+1} \leftarrow \mathbf{w} - U_{j+1}\mathbf{h}$;

(5) $H_{j+1} \leftarrow (\widehat{H}_j, \mathbf{h})$;

End For

Fig. 2.5 Algorithm: The k-step Arnoldi Factorization.

vides a means to determine a subset of the eigensystem. Hence, the ARPACK interface allows the user to specify the number l of eigenvalues sought.

When the Matrix A is considered in the Arnoldi method its l largest eigenvalues are determined. But in the case of the present study the interest lies in specific eigenvalues, e.g. those in the diurnal and semidiurnal spectrum. Therefore the shifted and inverted problem $(A - \sigma_0 I)^{-1}$ is considered. Thus the convergence of eigenvalues near the selected point σ_0 is enhanced. This approach is closely related to the inverse iteration techniques (e.g. [13]). Considering this spectral transformation in detail yields

$$A\mathbf{x} = \mathbf{x}\lambda \iff (A - \sigma_0 I)\mathbf{x} = \mathbf{x}(\lambda - \sigma_0). \tag{2.47}$$

and

$$(A - \sigma_0 I)^{-1}\mathbf{x} = \mathbf{x}\nu, \text{ where } \nu = \frac{1}{\lambda - \sigma_0}. \tag{2.48}$$

Hence, the eigenvalues λ that are close to σ_0 will be transformed into eigenvalues $\nu = \frac{1}{\lambda - \sigma_0}$, which are at the extremes of the transformed spectrum. The corresponding eigenvectors remain unchanged.

In case of the shifted and inverted approach of the Arnoldi Method, linear systems of the form $(A - \sigma_0 I)\mathbf{x} = b$ have to be solved. The algorithms of ARPACK are provided with a so called *reverse communication interface*. This interface allows the user to transfer the solution $\mathbf{x}$ into the algorithm, and in this way the solver can be chosen independently from ARPACK. In the present study the LU-solver provided by ScaLAPACK [4] is used (see next section). The LU-solver puts itself forward since the time consuming LU-factorization of $(A - \sigma_0 I)$ need to be performed only once.

2.2.2 The Parallelization with MPI

To enable the use of routines of mathematical libraries for computing linear systems, it is necessary to store the complex matrix $(A - \sigma_0 I)$ in a general form. Thus the advantages of the symmetries of the matrix are getting lost. Since more than 500 GB of memory are required, it is necessary to parallelize the ocean model and distribute the matrix on different nodes. The parallization is done with MPI[3], perfect for large problems needing access to large amounts of memory on distributed memory architectures [46].

The linear systems are solved with a parallelized version of a LU-solver of the ScaLAPACK software package [4]. Since the Matrix $(A - \sigma_0 I)$ is kept preserved during the whole iteration process of the Arnoldi algorithm, the LU-factorization, the most time consuming part, is only performed once. The choice of MPI and the ScaLAPACK LU-solver gives the user a high degree of freedom, in adapting the ocean model to the features of the computer architecture. The number of CPUs and nodes can freely be chosen, and is only restricted through the memory used to store the matrix.

2.2.3 The Performance of the Model

The model-runs have been performed on two distinct supercomputers, the HLRE[4] and the HLRS[5], equipped with NEC SX-6 nodes and NEC SX-8 nodes, respectively.

The number of free oscillations sought is set to $l = 150$ for each model-run. So

Table 2.2 Data of the performance of the fastest model-run on 12 NEC SX-6 nodes of the HLRE: First two rows shows values from one single cpu; last row are mean global values of all 96 cpus.

	Frequency	Time in [s]	Performance in [MFlops]
LU-factorization (of cpu no. 1)	1	6872.6	6766.8
LU-solver (of cpu no. 1)	500	444.7	1373.1
Mean global values of 96 cpus		8181.2	$608.7 \cdot 10^3$

---

[3] Message-Passing Interface.

[4] HLRE - High Perfomance Computing Centre for Earth System Research, Hamburg.

[5] HLRS - High Performance Computing Centre, Stuttgart.

75 free oscillations and the corresponding complex conjugated ones are computed. At the HLRE it is possible to run the program on 12 and 16 nodes. Each SX-6 is equipped with 8 cpus. The overall performance is up to 609 and 632 GFlops, respectively, being one of the fastest single-application running on the HLRE. The computation time is between 2 and 3 hours, mostly depending on the actual state of the supercomputer and on the condition of the matrix, which changes through changing σ_0 value. Although the LU-factorization is highly optimized (Table 2.2), it alone needs more than two third of the total time used by the model (in some cases up to 80%), the LU-solver uses 5-8%. The total memory of the model amounts to 630GBytes.

Furthermore, model-runs have been performed on the HLRS supercomputer. It is

Table 2.3 Data of the performance of model-runs on 4, 8, 16, 32 and 64 NEC SX-8 nodes of the HLRS.

Number of nodes	4	8	16	32	64
Number of cpus	32	64	128	256	512
Real Time in $[s]$	11421	6850	4463	3251	2766
Performance in $[GFlops]$	416	737	1269	2108	3394

one of the TOP 100 Supercomputers of the world[6], and ranked 48th in July 2006[7]. The model was distributed on up to 512 cpus (64 nodes). On this computer architecture the good performance of the model is kept preserved (Table 2.3), using only 45 minutes to determine 150 normal modes, with a mean performance of 3.4TFlops.

---

[6] http://www.top500.org/.
[7] The date when these model-runs have been performed.

Chapter 3
The Free Oscillations

The free oscillations of the World Ocean on the rotating Earth consist of gravity and vorticity modes, primarily governed by the gravity of the Earth and by the latitude dependent Coriolis force, respectively. A total of 284 free oscillations is found in the period range of 7.7 h to 165 h. Of these, 169 oscillations are classified as physically relevant, while the remaining 115 are considered to be spurious modes. These spurious modes are characterized by very large amplitudes in both the horizontal mass transport and the sea surface elevation. They are concentrated on small areas and located mainly in bays and gulfs. Although resonances do exist in these regions, they cannot be resolved with this model. In the following, only the physically meaningful modes are considered.

All 56 modes, computed by ZAMU2005 with the parameterized LSA-term and periods between 8.03 h and 133.10 h, correspond to modes that have now been computed with consideration of the full LSA-effect. As expected, the three single modes with the full LSA-effect, computed with a special iteration procedure in ZAMU2005, and the four modes calculated with the Wielandt Method by [29], have been verified in this work with exact match (compare Section 2.2). Additionally, for an analysis of the full LSA-effect and not its parameterization , 107 normal modes with periods between 12h and 120h are computed without the LSA-effect.

The accuracy of all these normal modes is of the same order of magnitude as that of those found using the inverse iteration method by ZAMU2005. The residual r ranges from 1.24E-13 to 1.23E-10 for the calculated eigensolutions (equation 2.46), corresponding to a correctness of the first six digits of the mantissa of σ_k.

3.1 Gravitational Modes

72 of the 169 determined normal modes are gravitational modes . The period of the slowest gravitational mode is 79.18 hours and the computed spectrum covers the period range down to 7.7 hours. In Table A.1 selected features of the gravitational

modes of the whole spectra are listed and Figures 5.1- 5.18 show the patterns of sea
surface elevation and energy flux of some of these modes.

3.1.1 The β-value

An approximation of the full integral formulation of the LSA-effect through a scalar
relationship (β) to the local tidal elevation, was introduced by [1]. Investigations of
this approximation in barotropic tidal models gave varying values of β. The value
of β depends on the tidal constituents [33] and differs between the open ocean and
the coastal regions [41]. Furthermore, [48] gave a polynomial dependency of a 'lo-
calized' β_L on the ocean depth D. (see also Section 2.1.4). In the present work, the
depth dependent $\beta_L(D)$ is determined for a few modes, that are relevant for semidiurnal and diurnal tidal dynamics (Figure 3.1). The results show a separation between
lower and larger β_L at roughly 2500m depth, similar to the polynomial of [48], but
with a dependency of β_L on the time scale. The curves are shifted towards larger
values for the modes with longer periods. The 26.20-mode most closely follows the
polynomial.

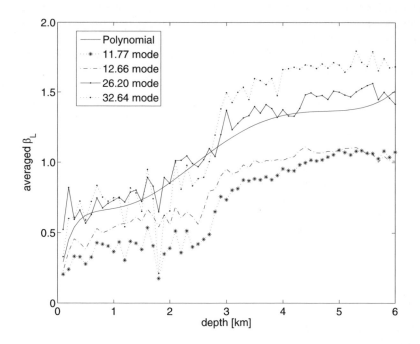

Fig. 3.1 Depth dependent β_L(D) determined for 4 modes. The solid line represents the polynomial
calculated by [48].

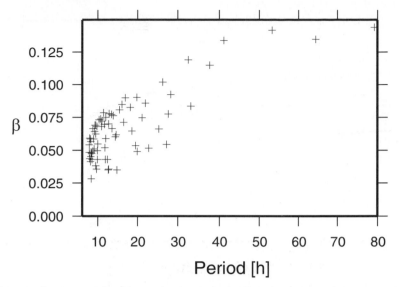

Fig. 3.2 The β-value, as defined by [1], for the gravitational modes with periods between 8 and 80 hours.

Calculations of the β-value, as defined by [1], show an increasing trend with increasing periods, as well (Figure 3.2). For the fast modes with periods shorter than 10 h, the mean β is 0.052 increasing to 0.065 for the semidiurnal (10-16 h) range. For the modes in the diurnal range (16-34 h), the averaged value is 0.076 and the slowest modes with periods longer than 34 h have β-values between 0.115 and 0.144.

3.1.2 The Influence of the LSA

Frequency

To analyze the direct effect of LSA on the gravitational modes, 51 corresponding modes neglecting the LSA-effect have been computed in the period range between 9.42 and 79.18 hours: The prolongation of the periods due to the LSA-effect appears in all modes, consistent with the theoretical estimate of [23]. The quality of this estimation is shown in Figure 3.3. The mean change in the period is 0.76 h (or 4%) and the largest period shift is 3.3 h for the mode with the period of 53.21 h. The changes in the decay time induced by the LSA-effect can have both positive and negative sign and the averaged variation of the respective decay times amounts to 7.4%.

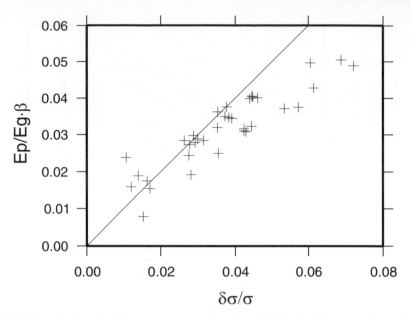

Fig. 3.3 The ratio of potential to total energy E_p/E_g times the β-value of modes with calculated without LSA in dependency of the relative frequency shift induced through the LSA-effect. Thin line: theoretical approach (equation 2.36); +: 35 gravity modes in the period range 12 to 80 hours.

The spatial pattern

The spatial patterns of the sea surface elevation ζ and the horizontal mass transport **u** are likewise changed noticeably through the LSA-effect. The main patterns of the sea surface elevation and mass transport ellipses are preserved. But changes occur in the strength of local resonances and in slight shifts of the amphidromies.

Examination of the energy contents of the different parts of the ocean (Indic, Pacific, Atlantic, Southern Ocean, Arctic Ocean) show mean energy changes of 1.8%, 5.1%, 3.9%, 1.2% and 0.7%, respectively, of the total energy E_t. However, a few modes show variations of up to 6.5%, 43.3%, 37.0%, 6.9% and 8.7%, due to large changes of local resonances.

The 12.66-mode and the 12.76-mode (Figures 3.4) are both excited by the semidiurnal tides; indeed, the 12.66-mode is the major component of the M_2-tide (see Section 4.1). While these two modes have similar structures in their spatial patterns of ζ and **u**, the 12.66-mode is dominated by the resonance in the North and Equatorial Atlantic region and the 12.76-mode has its main energy in the Pacific. In PL1981, this resonance in the Atlantic appears in mode 33 (12.1h) and the equatorial transverse 3/2 wave in the Pacific in mode 35 (11.6h). In the Atlantic Ocean northward of $30°S$, the energy amounts to 40% of the total energy (E_t) for the 12.66-mode and to 13% for the 12.76-mode. The corresponding modes calculated without the LSA-term (Figure 3.4) have periods shortened by 0.5h, and the resonances in the Atlantic are significantly underestimated (12% of E_t) for the faster and overes-

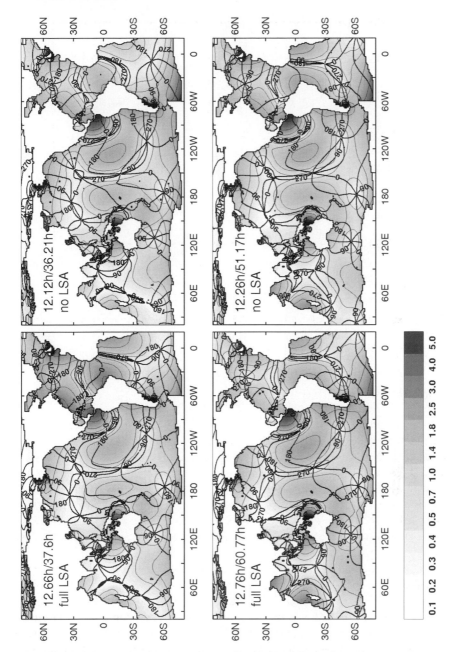

Fig. 3.4 Normalized amplitudes of sea-surface elevation and lines of equal phases. **Top**: The 12.66-mode with a decay time of 37.60 h and the corresponding mode calculated without the LSA-effect with a period of 12.12 h and a decay time of 36.21 h. **Bottom**: The 12.76-mode with a decay time of 60.77 h and the corresponding mode calculated without the LSA-effect with a period of 12.26 h and a decay time of 51.17 h.

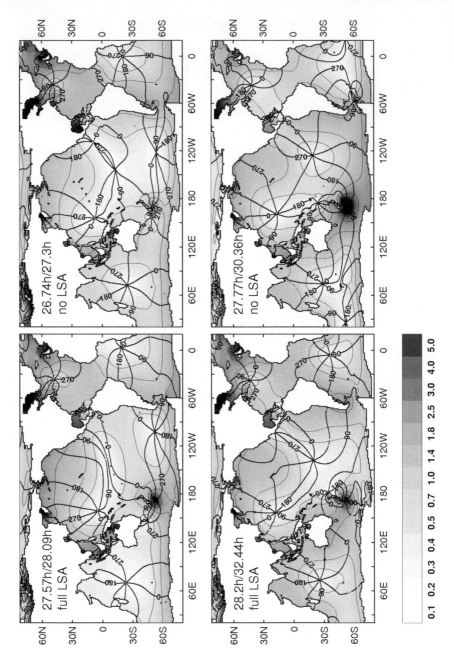

Fig. 3.5 Normalized amplitudes of sea-surface elevation and lines of equal phases. **Top**: The 27.57-mode with a decay time of 28.09 h and the corresponding mode calculated without the LSA-effect with a period of 26.74 h and a decay time of 27.30 h. **Bottom**: The 28.20-mode with a decay time of 32.44 h and the corresponding mode calculated without the LSA-effect with a period of 27.77 h and a decay time of 30.36 h.

timated (21% of E_t) for the slower one. Another interesting effect is represented by the two corresponding modes computed with parameterized LSA-term (12.65- and 12.75-mode, ZAMU2005). Both have almost the same periods as the ones computed with full LSA-term, but the characteristics of the resonances are similar to those computed without LSA-term.

Two other modes for which the LSA-effect introduces large changes to local resonances are the 27.57- and the 28.20-mode (Figure 3.5). These are two gravitational modes affected by a topographical vorticity mode, which is located at the New Zealand Plateau . This vorticity-wave-trapping seems to be subject to the same mechanism, analyzed by PL1981. The vorticity mode is concentrated at the New Zealand Plateau where two topographical vorticity modes occur with periods 32.56 hours (Figure 5.20) and 37.92 hours. The iteration method used by ZAMU2005 did not determine any topographical vorticity mode in this area and the fastest of PL1981 is mode 11 (38.0 h). Probably the faster mode is not resolved in PL1981 due to the lower spatial resolution. Hence, the vorticity-wave-trapping occurs at this location for gravitational modes with longer periods in the modes of PL1981 than for those in the present study. The strength of the trapping within the two gravitational modes (27.57- and 28.20-mode) mentioned above is very sensitive to the LSA-effect. It can be quantitatively expressed by the variation of the ratio of potential to total energy $\frac{E_p}{E_t}$, since the vorticity modes themselves have very low ratios of around 2%. Both modes computed with the LSA-effect have a ratio between 26% and 27% and they are strongly influenced through this vorticity mode. In contrast, neglecting the LSA-term results in a reduced ($\frac{E_p}{E_t}$=39%) for the faster mode (26.74 h) and in a strengthened influence ($\frac{E_p}{E_t}$=9%) for the slower mode (27.77 h). For the corresponding pair computed with parameterized LSA-term (27.66- and 28.15-mode, ZAMU2005), both the frequency and $\frac{E_p}{E_t}$ (between 23 and 25%) are similar to that of the modes computed with the full LSA-term.

The 32.64-mode has a first-order Antarctic Kelvin Wave (AKW1) structure (see following section) and is the most important mode for the diurnal tides (ZAMU2005) and for atmospheric pressure forcing (compare the EOF for the 30-36 h period band [39]. The LSA affects mainly the period with a prolongation of 2.4 h, and the spatial patterns of ζ and $\mathbf{u}$ are only slightly changed. Altogether three gravitational modes are characterized by an AKW1 (32.64, 33.25 and 37.77 h) and all three modes are affected by topographical vorticity modes at the Kerguelen and the New Zealand Plateaus. The strength of this trapping is not sensitive to the LSA-effect. Figure 5.15 gives a graphical display of the 32.64-mode. Vorticity-wave-trapping is indicated by distortions of the phase lines of the sea surface elevation in these regions. PL1981 already showed that one of their AKW1s is affected by a resonance at the Kerguelen Plateau (Mode 15). [53] also describe an interaction between the AKW1 and topographic modes. However, the trapping of a vorticity mode at the New Zealand Plateau has not yet been noted, although it is clearly present in the patterns of the diurnal tides [63, 51]).

3.1.3 The Antarctic Kelvin Wave

A prominent feature appearing in the Southern Ocean is the Antarctic Kelvin Wave. This wave travels in westward direction trapped by the Antarctic and and it is accompanied by an strong westward energy flux. The Antarctic Kelvin Waves have orders of one, two and three (Figure 5.19), defined through the number of wavelengths within one cycle.

The Kelvin Waves with order one have periods of 32.64, 33.25 and 37.77 hours. The 32.64-mode (5.15), is an important mode for the diurnal tides although it the period is not close to the forcing periods. This is due to the large coherence of the diurnal potential with the adjoint mode of the 32.64-mode (Section 4.1). Along the path of the first order waves topographical trapping occurs at the Kerguelen Plateau and the New Zealand Plateau (Section 3.1.2). The second order Antarctic Kelvin Waves have periods of around 16 hours (16.02-mode and 16.89-mode, Figure 5.10). These modes are neither important for the diurnal nor for the semidiurnal tides. The third order waves are in the semidiurnal period range (10.98-mode and 11.65-mode, Figure 5.3) and the 11.65 is significantly excited through the semidiurnal tidal potential.

3.1.4 New Modes

Several new gravitational modes having not been captured by the iteration method of ZAMU2005 are computed by means of the Arnoldi Method. Some of them with periods in the semidiurnal and diurnal period range are described below:

Semidiurnal

The 11.98-mode (Figure 5.5) has its main energy located in the Pacific. There, it appears mainly as a Kelvin Wave around New Zealand, which is a prominent feature of the semidiurnal tides and indeed this mode plays an important role in the synthesis of the semidiurnal tides (Section 4.1). The 11.98-mode resembles the mode 38 (10.8 h, Fig.20) of PL1981 who named it the "free New Zealand Kelvin Wave".

Another new mode which plays an interesting role in the synthesis of the semidiurnal tides is the 13.49-mode (Figure 5.8). Almost all energy is located in the Pacific (92.5%), represented by a Kelvin Wave travelling along the South and North American coast, ending at the Aleutian Island. The patterns of this mode are very similar to the 13.37-mode and both modes are strongly excited through the semidiurnal tidal force. However, since they are forced with opposite phases they combine destructively on a global scale (Section 4.1).

Diurnal

The 25.32- (Figure 5.13) and 27.19-mode (Figure 5.14) are important for the diurnal tides since their periods are close to that of the diurnal tidal forcing. Both are restricted to the Pacific (more than 98% of E_t) and characterized by a single amphidrome. The latter is located at the equator and forms a Kelvin Wave along the northern Pacific coast with strong resonances in the Sea of Okhotsk and the South China Sea. There are no counterparts found in PL1981.

3.1.5 The Slowest Modes

The four slowest modes are the 41.22-, 53.21-, 64.36- and 79.18-mode (Figure 5.17 and 5.18). The ratio of potential to total Energy $\frac{E_p}{E_t}$ decreases from 42.2% down to 18.6% with increasing period of the modes. Hence, the rotational character gains the more influence the slower the mode is. All four modes are characterized by an eastward directed belt of energy flux surrounding the world. The 53.21-mode has its main energy in the Atlantic (45.7% of E_t). The North and South Atlantic are co-oscillating where the sea surface elevations showing a separation through a quasi-nodal line at the equator. Thus each ocean plays the role of a Helmholtz resonator with respect to the other. The same phenomenon is analyzed by PL1981 but for a mode with the somewhat lower period of 35.8 h (mode 12). Furthermore, for the 53.21-mode in the North Atlantic a strong topographical trapping at the Reykjanes Ridge occurs.

The two slowest gravitational modes, the 64.36- and the 79.18-mode, have been classified as gravitational modes since they would not disappear on a non-rotating Earth. However, when allowing for earth rotation they have features of vorticity modes indicated by the low ratio of potential energy to kinetic energy. The sea surface elevations have nearly constant amplitude and phases in the North Atlantic, similarly to the sea surface elevation patterns of some planetary modes (Section 3.2.2). Furthermore, the sea surface elevations in the South Atlantic, the Pacific and the Indic are similar to those of planetary modes. Structurally the 79.18-mode resembles mode 1 (79.8 h) of PL1981. However, it shows a uniform energy distribution with a circum-global energy flux and only a weak topographical trapping at the Falkland Plateau. In the Pacific, it is similar to the fundamental planetary-topographic mode of MI1989 (74.4 h, see his Fig.10). Thus, these results suggest that this mode (or even the 64.36-mode) could also be classified as the first planetary mode of the World Ocean.

3.2 Vorticity Modes

3.2.1 Topographical Vorticity Modes

In total, 91 of the 97 determined vorticity modes are topographical modes . They are trapped by prominent topographic structures over which they focus the main part of their total energy. They are characterized by an energy-flux gyre (see also PL1981 mode 14) and a strong modification of their relative vorticity with changing water depth (more precisely: normal to the contours of $\frac{f}{h}$). The behavior of the transport ellipses is characterized by an inversion of the sense of rotation at a certain depth contour line (e.g. Figure 5.21). The rotation is anti-cyclonic over the topographic structure and cyclonic in deeper water, consistent with previous observations and numerical models (e.g. [18]).

In consideration of the fact that these resonances occur in specific areas, they are indexed by 12 different regions (Figure 3.6). Note, it is often the case that these free oscillations comprehend resonances in more than one of these regions.

For the *Arctic Ocean*, comprising extensive shelf areas and the transition to the Arctic Basin, the vorticity mode with the lowest period (13.74 h) exists. Altogether 15 modes have resonances in this region and are distributed to a broad period range up to 101.22 h. Almost the whole spectrum of the calculated topographical modes is covered by modes occurring in the region around *New Zealand* with the New Zealand Plateau , the Lord Howe Rise, the Colville Ridge and the Fiji Plateau (Figure 5.20), the *Pacific Antarctic Ridge* and the *Kerguelen Plateau* (Figure 5.22). The oscillations with their center located on the *Falkland Plateau* are in a similar range between 27.18 and 92.47 hours. Under *North Atlantic*, resonances in the Norwegian Basin, at the Island-Faroe Ridge, at the Rockall Plateau and at the Grand Banks of New Foundland are summarized. These resonances are limited to low periods between 28.08 and 47.58 hours like those occurring in the *Ross Sea* and in the *Weddell Sea* with periods between 30.13 and 40.24 hours. The *Mid Atlantic Ridge* is defined by the part to the north of $20°N$ including the Reykjanes Ridge (Figure 5.21) and the part to the south of $20°S$ (Figure 5.23). For resonances at the northern ridge system the periods are widely spread from 51.78 to 132.19 hours. Since the southern part is connected with the *South-West Indian Ridge*, some modes with periods over 130 hours are trapped by both topographical structures, and for the modes in both regions a lower limit of around 80 hours exists. Oscillations with the focus of their energy at the *South-East Indian Ridge* often extend far into the Indian Ocean including the *Central Indian Ridge*. They have in any case periods larger than 105.74 hours and are densely distributed up to 153.20 hours. Only a few modes do not fall under this regional classification scheme, they occur at the Mariana Islands (106.06 h), the Bering Sea (33.09 h), and the Gulf of Alaska (125.99 h, 129.72 h).

To analyze the effect of LSA on topographical modes, a set of 58 corresponding modes neglecting the effect have been computed. The LSA-effect lengthens the periods of these modes by a mean shift of 0.17 h. The small changes, compared to the long periods, are in good agreement with the theoretical estimates of [23],

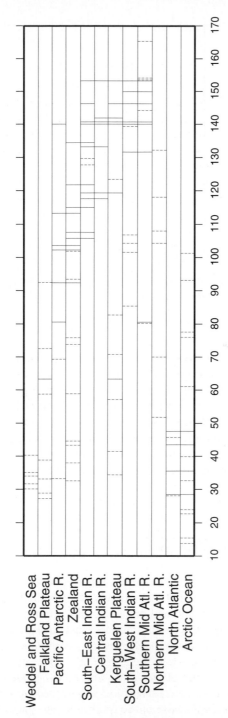

Fig. 3.6 The period spectra of the topographically controlled vorticity modes in the period range 13-165 hours, subdivided into 12 main regions. Solid line: mode with resonances in more than one of these regions; dashed line: mode with a resonance in exactly one of these regions.

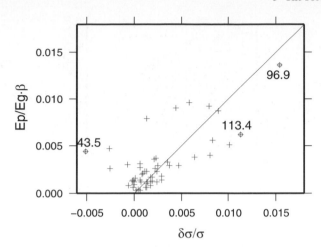

Fig. 3.7 The quantity $\frac{E_p}{E_t} \cdot \beta$ of modes with LSA neglected, in dependency on the relative frequency shift induced by the LSA-effect. Thin line: theoretical approach (equation 2.36); +: 58 vorticity modes modes in the period range 13 to 120 hours, 2 planetary vorticity modes (Periods: 96.9 h and 113.4 h) and a topographical vorticity mode (Period 43.5 h) are marked by **o** (see text for explanations).

since the ratio $\frac{E_p}{E_t}$ for all these modes is small (Figure 3.7). However, 11 modes have a negative shift in the period, which cannot be explained by the theoretical approach of [23]. Topographic roughness (or small-scale topographic features) may explain this shift towards higher frequencies via second-order perturbation theory as explained by MI1989. These shifts, however, are quite small, with only three being larger than one thousandth of its period. The largest of these anomalous shifts, 0.22 h, occurs for the mode with a period of 43.49 h (Figure 5.21). It is centered at the Faeroe-Island Ridge and it resembles the mode that was measured and analyzed by [28].

3.2.2 Planetary Vorticity Modes

Altogether six planetary modes have been computed with periods of 96.94, 111.22, 113.44, 119.65, 136.87, and 141.66 h (Figures 5.25-5.27) and Table 3.1). Five of them have a global distribution and one (113.44h) is limited to the Pacific Ocean. They have been classified as planetary modes since they would exist in the World Ocean with constant depth, as well. Setting the ocean depth to 4000 m, for all of these modes a corresponding one is found. The sea surface elevation and the energy flux of the descendants of the 119.65-mode and 136.87-mode are shown in Figure (3.9). They have periods of 123.21 h and 137.96 h, respectively. The spatial patterns of ζ and **u** are smoother than those of the original mode, since all topographical effects are extinguished.

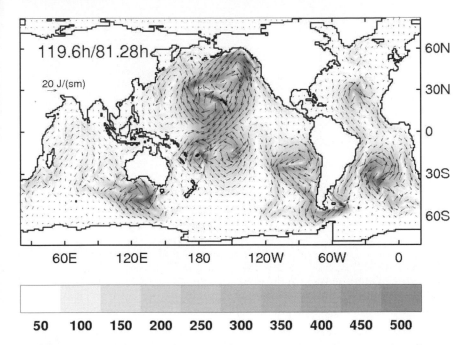

Fig. 3.8 Planetary 119.65-mode with a decay time of 81.28 h. Energy flux vectors shown for every fifth grid point, zonally and meriodionally. Squaring the magnitudes given yields these quantities in J/(sm). The energy flux vector at the top left has a magnitude of 20^2 J/(sm). The magnitude of the energy flux is additionally gray shaded.

For the modes with periods of 96.94 h and 113.44 h, moreover the corresponding modes without LSA-effect have been computed. An effect of the LSA on the two fastest planetary modes are positive period shifts of 1.5h and 1.3h, which are larger than the shifts of topographical modes (Figure 3.7). Similar magnitudes of the frequency shifts for the other four planetary modes are expected, because the β-values for all of these modes are quite large (between 0.11 and 0.14), and the ratios of E_p/E_t are larger than 5% (compare [23]). This analysis confirms the statement of [48], having shown that the LSA-effect on atmospherically forced (2 to 7 days) barotropic dynamics is not negligible. However, the effect of LSA on the spatial patterns of the sea surface elevation and the horizontal mass transports is negligibly small for both, topographical and planetary vorticity modes.

The planetary modes are relevant to atmospherically forced barotropic dynamics (e.g., [26, 38]).

The 96.94-mode (Figure 5.25) has counterparts in the 96 h vorticity mode of PL1981, calculated in an ocean with uniform depth (see their Fig.8) and in the 96.32-mode of ZAMU2005 (see their Fig.11) when allowing for parameterized LSA. The structural differences between the parameterized or full LSA modes, are negligible.

The structure of the 111.22-mode (Figure 5.25) is very similar to that of the mode

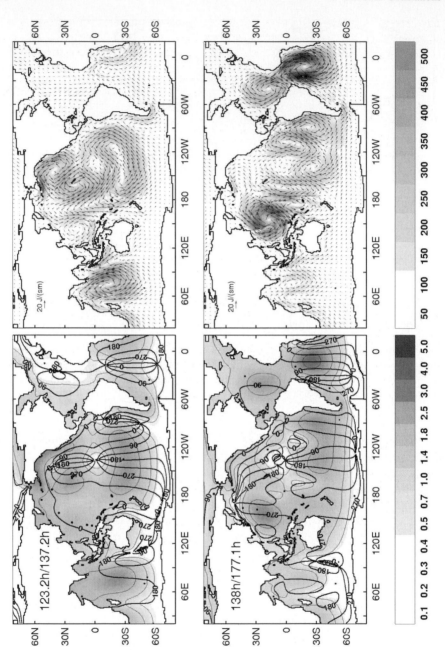

Fig. 3.9 Planetary Modes, calculated with constant depth (D=4000 m). **Left**: Normalized amplitudes of sea-surface elevation and lines of equal phases in degrees referred to 0° in (0.5°N, 89.5°W) (in steps of 45°). **Right**: Energy flux vectors shown for every fifth grid point, zonally and meriodionally. Squaring the magnitudes given yields these quantities in J/(sm). The energy flux vector at the top left has a magnitude of 20^2 J/(sm). Further, the magnitude of the energy flux is gray shaded. **Top**: The 123.21-mode with a decay time of 137.19 h. **Bottom**: The 137.96-mode with a decay time of 177.11 h.

Table 3.1 The six planetary modes in the period range 96 to 142 hours. The period $T_2 = \frac{2\pi}{\bar{\sigma}_2}$, the decay time $T_1 = \frac{1}{2\bar{\sigma}_1}$, the period T_2^p of the corresponding mode computed with parameterized LSA (if computed), and the ratio between potential (E_p) and total (E_t) energy contents are shown for each mode. Further, areas relative to the ocean area are given in per cent for Indian Ocean (Ind), Pacific Ocean (Pac), Atlantic Ocean (Atl), Southern Ocean (Ant) and the North Polar Sea (Np) directly below the abbreviations. The five columns below these abbreviations show the total energy contents of the corresponding ocean region relative to the total energy. The last column shows the β-value.

No.	$T_2[h]$	$T_2^p[h]$	$T_1[h]$	$\frac{E_p}{E_t}[\%]$	Ind 18.7	Pac 46.6	Atl 23.2	Ant 8.5	Np 3.0	β
1	96.94	96.32	88.64	11.63	31.09	41.08	25.03	2.78	0.02	0.140
2	111.22	110.83	87.58	8.13	10.43	23.74	64.18	1.55	0.10	0.119
3	113.44		78.14	6.46	0.49	96.75	1.97	0.79	0.00	0.125
4	119.65	119.42	81.28	7.67	11.26	66.82	19.86	1.99	0.06	0.132
5	136.87		80.89	5.10	10.08	80.39	6.68	2.85	0.01	0.112
6	141.66		78.73	5.57	7.45	55.51	34.91	2.11	0.01	0.116

of [15] (see their Fig.1, 120 h). The main part of the energy is located in the Atlantic (62% of E_t). The sea surface elevation has nearly constant amplitude and phase in the North Atlantic and a strong resonance in the South Atlantic. A resonance in the Southeast Pacific (described later in more detail for the 119.65-mode) transports energy through the Drake Passsage to the South Atlantic. The Indian Ocean has constant amplitude and phase values over large regions.

The third planetary mode (113.44 h, Figure 5.26) is restricted to the Pacific and has its main energy in the northern part, where it appears as a westward propagating planetary wave with one wavelength. Topographical trapping occurs at the north of the Fiji Islands. The cross equatorial structure of the propagating planetary wave and the amphidromic point at the Gulf of Alaska suggest that this mode is a descendant of the mode of MI1989 (see his Fig.11c) with a period of 132 h.

The planetary mode with a period of 119.65 h (Figure 5.26) appears in ZAMU2005 with a period of 119.42 h (see their Fig.12). It has a global energy distribution with 13% of E_t in the Indic and Southern Ocean through topographical trapping at the Southeast Indian Ridge. Of the total energy, 20% is located in the Atlantic, mainly in a resonance in the middle of the South Atlantic, and 67% in the Pacific. The structure in the Pacific is similar to that of MI1989 (see his Fig.4d) with a period of 120 h, but it appears in the 119.65-mode meridionally compressed and bounded in the east by the East Pacific Rise. An additional resonance east of the East Pacific Rise arises from a planetary wave with half wavelength. The energy flux (Figure 3.8) shows that the ridge reflects the energy and, thus, is responsible for this subbasin (bounded in the east by South America) resonance. Part of this energy is transported through the Drake Passage in the Atlantic. It is interesting that the phase and amplitude structure in the South Atlantic, and the amplitude structure in the Southeast Pacific, appear in the sea surface elevations of the atmospherically forced ocean model of [38] (see his Fig.3). But for the dynamical signal (see his Fig.4) this structure disappears.

In the 136.87-mode (Figure 5.27), 80% of E_t is located in the Pacific. In this mode, strong topographical-trapping components exist. The strongest is in the southern part at the Pacific-Antarctic Ridge, in the north-west part at the South Honshu Ridge and a weak trapping at the Southeast Indian Ridge. In the Pacific this mode may correspond to the mode of MI1989 with a period of 139.2 h (see his Fig.11d).

The slowest computed planetary mode has a period of 141.66 h (Figure 5.27). The main energy is located in a strong resonance in the subbasin bounded by the East Pacific Rise transporting energy through the Drake Passage in the middle of the South Atlantic. For this region, the spatial structure is similar to that of the 119.65-mode, but more energy is located in this region.

Recapitulating, we find that the Atlantic Ocean, with its connection to the Pacific through the Drake Passage, plays an important role in the formation of all global planetary modes. Furthermore, we see that some of these modes are strongly influenced by the bottom topography, either through trapping or by functioning as a barrier.

Chapter 4
Synthesis of Forced Oscillations

The semidiurnal and diurnal tidal oscillation systems of the open ocean are meanwhile well known. This is due to many improvements made during the last decades. On the one hand, missing physical effects were included, e.g. the LSA-effect (e.g. [1, 56]) and the parameterization of the internal wave drag [7]. On the other hand, the assimilation of tidal sea surface elevations extracted from satellite altimetry (TOPEX/POSEIDON) brought about very accurate tidal solutions (e.g. [63]) .

To understand the tidal oscillation system, the tidal solution can be represented spectrally. This attempt describing ocean tides by synthesizing free oscillations, was first made by [35]. His free oscillations were obtained by solving the homogeneous Laplace tidal equation without frictional terms and loading and self-attraction effects (PL1981). Dissipation was included into the synthesis with a special procedure [34]. The large scale features of the resulting synthesized solutions were in good agreement with complete tidal model solutions. Hence, further investigations of the spectral composition of the semidiurnal and diurnal tides were made [35].

Another spectral composition was made by ZAMU2005. They used a simple least squares approximation to determine the contribution of individual free oscillations to the tidal patterns.

In this chapter a general procedure of the synthesis of forced oscillations is derived (Section 4.1.1). The set of free oscillations is computed with explicit consideration of frictional terms and the full LSA-effect (Chapter 2). For the synthesis, the solution of the corresponding adjoint eigenproblem is necessary. The synthesis procedure allows for an analysis of the strength and the phase of each normal mode in respect to each semidiurnal and diurnal tidal constituent. Herewith, the LSA-effect on the tidal dynamics is investigated (Section 4.1.2), explaining, e.g. the phase delay induced by the LSA in tidal models and the role of destructive interference of certain free oscillations. Furthermore, following the work of [35], various features of some semidiurnal and diurnal tidal oscillation systems are spectrally analyzed (Section 4.1.3).

4.1 Tidal Dynamics and the Influence of LSA

4.1.1 The Procedure of Tidal Synthesis

Theory

The tidal equations are written in the general operator notation;

$$\frac{\partial \mathbf{w}}{\partial t} + \mathscr{L}\mathbf{w} = \tilde{\mathbf{F}}$$
$$\mathbf{w} = (\zeta, u, v), \tag{4.1}$$

where ζ is the sea surface elevation and $\mathbf{u} = (u, v)$ is the horizontal velocity vector. $\tilde{\mathbf{F}}$ is the external forcing term and the operator $\mathscr{L}$ represents the tidal dynamics. The decomposition of $\tilde{\mathbf{F}}$ into its spectral constituents $\mathbf{F}e^{-i\sigma_F t}$ with frequency σ_F yields:

$$(\mathscr{L} - i\sigma_F)\mathbf{w}_F = \mathbf{F} \tag{4.2}$$

with the partial tide $\mathbf{w}'_F = \mathbf{w}_F e^{-i\sigma_F t}$. Forced tidal oscillations $\mathbf{w}_F$ can be expressed through a superposition of free oscillations $\mathbf{v}_k$

$$\mathbf{w}_F = \sum_{k=1}^{\infty} a_k \cdot \mathbf{v}_k. \tag{4.3}$$

The a_k are the expansion coefficients and $\mathbf{v}'_k = \mathbf{v}_k e^{-i\sigma_k t}$ are the eigenfunctions, defined through the homogeneous equation

$$(\mathscr{L} - i\sigma_k)\mathbf{v}_k = 0. \tag{4.4}$$

The eigenfrequencies σ_k are complex valued, $i\sigma_k = \sigma_{k,1} + i\sigma_{k,2}$, with the oscillatory $\sigma_{k,2}$ and the damping part $\sigma_{k,1}$.

The determination of the expansion coefficients in (4.3) requires knowledge of the eigenvectors $\widehat{\mathbf{v}}_k$ of the adjoint eigenproblem:

$$(\widehat{\mathscr{L} - i\sigma_k})\widehat{\mathbf{v}}_k = 0. \tag{4.5}$$

In case of no dissipation the [?] $\widehat{\mathbf{v}}_k$ are the complex conjugated of the corresponding eigenvectors $\mathbf{v}_k$. The addition of dissipation makes the operator $\mathscr{L}$ non self-adjoint and thus the [?]s have to be determined separately.

With an arbitrary scalarproduct $\langle \rangle$, the adjoint $\widehat{O}$ of an operator O is defined by $\langle O\mathbf{w}, \overline{\mathbf{w}} \rangle = \langle \mathbf{w}, \widehat{O}\overline{\mathbf{w}} \rangle$, $\mathbf{w}$ and $\overline{\mathbf{w}}$ denoting arbitrary vector functions. The eigensolutions of the two above mentioned eigenproblems (4.4) and (4.5) are forming biorthogonal systems of eigenfunctions $\{\mathbf{v}_k\}_{k=1,\infty}$ and $\{\widehat{\mathbf{v}}_k\}_{k=1,\infty}$. Properly normalized, these eigenfunctions satisfy the condition

$$0 = \langle \mathscr{L} \mathbf{v}_k, \widehat{\mathbf{v}_l} \rangle - \langle i\sigma_k \mathbf{v}_k, \widehat{\mathbf{v}_l} \rangle = \langle \mathbf{v}_k, \widehat{\hat{L}\mathbf{v}_l} \rangle - \langle i\sigma_k \mathbf{v}_k, \widehat{\mathbf{v}_l} \rangle =$$

$$\langle \mathbf{v}_k, \widehat{i\sigma_l \widehat{\mathbf{v}_l}} \rangle - \langle i\sigma_k \mathbf{v}_k, \widehat{\mathbf{v}_l} \rangle = \langle \mathbf{v}_k, \widehat{\mathbf{v}_l} \rangle (i\sigma_l - i\sigma_k)$$

$$\Rightarrow \langle \mathbf{v}_k, \widehat{\mathbf{v}_l} \rangle = \delta_{kl}, \qquad (4.6)$$

with the Kronecker symbol δ_{kl}.

Substituting (4.3) into (4.2) we obtain

$$(\mathscr{L} - i\sigma_F) \sum_{k=1}^{\infty} a_k \cdot \mathbf{v}_k = \mathbf{F}$$

$$\Rightarrow \langle \sum_{k=1}^{\infty} i(\sigma_k - \sigma_F) \cdot a_k \cdot \mathbf{v}_k, \widehat{\mathbf{v}_l} \rangle = \langle \mathbf{F}, \widehat{\mathbf{v}_l} \rangle, \qquad (4.7)$$

herein deriving the second relationship (4.4) is used and multiplied scalarly by an arbitrary adjoint eigenvector $\widehat{\mathbf{v}_l}$. Taking into account the orthogonality relation (4.6), we finally obtain for the expansion coefficient:

$$a_k = \frac{1}{i(\sigma_k - \sigma_F)} \cdot \langle \mathbf{F}, \widehat{\mathbf{v}_k} \rangle \qquad (4.8)$$

Thus, with the knowledge of the eigenvectors $\{\mathbf{v}_k\}_{k=1,\infty}$ and $\{\widehat{\mathbf{v}_k}\}_{k=1,\infty}$, and the related eigenvalues $\{\sigma_k\}_{k=1,\infty}$, every forced oscillation can be obtained by superposing. The magnitude and phase of the complex expansion coefficients a_k depend on the 'resonance depths' $R_k := \frac{1}{i(\sigma_k - \sigma_F)}$ and on the 'shape factor' $C_k := \langle \mathbf{F}, \widehat{\mathbf{v}_k} \rangle$. The latter reflect the spatial coherence of the forcing field with the adjoint eigenvectors. Both, the resonance depths and the shape factors, are determining the strength and phase of the excitation of the free oscillations.

The terms 'resonance depth' and 'shape factor' are chosen according to the notations of [23] and [36], respectively. Important to note is that these two earlier works discuss the tidal dynamics in terms of the Laplace tidal equation without dissipation and LSA-effects. This makes further simplifications possible and consequently, the formulae of the resonance depths and the shape factors differ from those derived above.

Model

The barotropic free oscillations of the World Ocean are computed with explicit consideration of dissipative terms and the full LSA-effect. The equation (4.4) is discretized with a finite difference model (Section 2.2), leading to the algebraic form

$$(A - i\sigma_k)\mathbf{x}_k = 0, \qquad (4.9)$$

with the NxN-Matrix A and the vector of unknowns $(\zeta, u, v) = \mathbf{x}$.
When defining the scalarproduct[1] by, $\langle x, y \rangle = \sum_{i=1}^{N} x_i \cdot y_i^*$, the adjoint of the matrix A
is the conjugate complex and transposed matrix, A^{T*}:

$$\langle Ax, y \rangle = \sum_{j=1}^{N} \sum_{i=1}^{N} a_{ij} x_i y_j^* = \langle x, A^{T*} y \rangle \tag{4.10}$$

where the $*$ denotes the conjugate complex. Finally, the adjoint equation can be
written as:

$$(A^{T*} + i\sigma_k^*)\widehat{\mathbf{x}}_k = 0. \tag{4.11}$$

Normally the scalarproduct of vectors which are defined on a global grid, is defined
as a Riemann sum and takes the area elements dA into account. The definition of the
scalarproduct affect the adjoint of the matrix A and thus the adjoint eigenvectors $\widehat{\mathbf{x}}_k$,
as well. However, the primary interest of this study is not in the adjoint eigenvectors
but in the expansion coefficients (4.8), which are independent of the choice of the
scalarproduct. For the analysis of the expansion coefficients it is important to choose
an appropriate normalization of the eigenvectors (see the following section).
The two eigenproblems of equation (4.9) and (4.11) are solved in the period range
from 9 to 40 hours. As expected, the eigenfrequencies of these two eigenproblems
are the same. The residual r of the eigensolutions of (4.9) and (4.11) is defined
through $r = \|A\mathbf{x}_k - \sigma_k \mathbf{x}_k\|_2$ and $r = \|A^{T*}\widehat{\mathbf{x}}_k + i\sigma_k^*\widehat{\mathbf{x}}_k\|_2$, respectively. The values of
r are lower than 1.3E-10, corresponding to a correctness of the first six digits of
the mantissa of σ_k [64]. Further, the accuracy of the orthogonality relation (4.6)
is essential in the derivation of equation (4.8) of the expansion coefficients. This
relation is fulfilled in the numerical model with an error lower than $err = 5 \cdot 10^{-7}$.

Analysis

The tidal synthesis is performed for eight tidal constituents of second degree, four
semidiurnal (K_2, S_2, M_2, N_2) and four diurnal (K_1, P_1, O_1, Q_1) ones. The tidal force
is (e.g. [3]):

$$\mathbf{F} = (0, F_u, F_v)$$

$$F_u = \frac{1}{R\cos\phi} \frac{\partial}{\partial \lambda} \Phi$$

$$F_v = \frac{1}{R} \frac{\partial}{\partial \phi} \Phi, \tag{4.12}$$

where the potential of second degree is defined as

$$\Phi(\lambda, \phi) = \gamma_2 K G_2^s(\phi) e^{is\lambda}, \tag{4.13}$$

---

[1] It is postulated that equation (4.9) is dimensionless, which is possible with a simple matrix transformation.

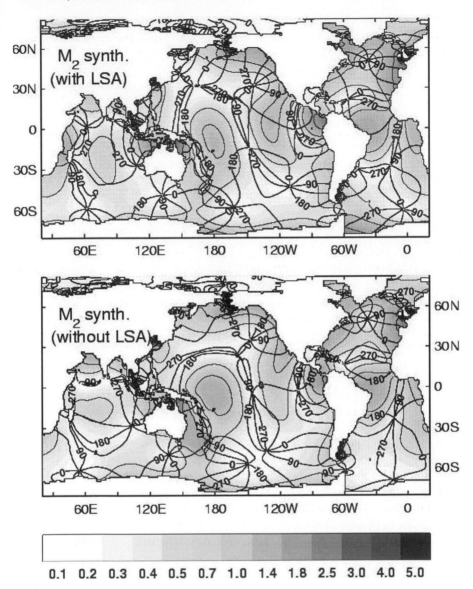

Fig. 4.1 Synthesized M_2-tide with (top) and without (bottom) consideration of the LSA effect: Gray shaded amplitudes of the sea surface elevation ζ in $[m]$. Solid lines showing the phases in degrees (in steps of $30°$).

with $\gamma_2 = 1 + k_2 - h_2$. There, k_2 and h_2 are the nonloading Love numbers, K is a specific coefficient for each partial tide, and $s = 2$ or 1 for the semidiurnal or diurnal potential, respectively. The geodetic functions $G_2^s(\phi)$ are given through the associated Legendre functions:

$$G_2^1(\phi) = sin2\phi$$
$$G_2^2(\phi) = cos^2\phi. \tag{4.14}$$

The tidal solution $\mathbf{w}_F$ of equation (4.2) would be exactly equal to the synthesized tide when using the complete spectrum of N eigenfunctions, but only the eigenfunctions within the semidiurnal and diurnal spectrum are used for the synthesis. Thus, the synthesized tide $\tilde{\mathbf{w}}_F$ differs from the tidal solution $\mathbf{w}_F$. However, as it is shown below, the difference between these two solutions is small.

In order to make the expansion coefficients a_k independent of scale factors of the tidal potentials, equation (4.3) is transformed, by using (4.8, 4.13), in:

$$\tilde{\mathbf{w}}_F = \sum a_k \cdot \mathbf{v}_k = \gamma_2 K \sum a'_k \cdot \mathbf{v}_k. \tag{4.15}$$

Furthermore, the eigenfunctions are normalized:

$$\bar{\mathbf{v}}_k = \frac{\mathbf{v}_k}{\sqrt{\int_S \zeta_k \cdot \zeta_k^* dA \cdot A_O^{-1}}} =: n_k \cdot \mathbf{v}_k, \tag{4.16}$$

where A_O is the area of the ocean domain. Finally, the synthesized tide can be written as

$$\tilde{\mathbf{w}}_F = \gamma_2 K \sum \frac{a'_k}{n_k} \cdot \bar{\mathbf{v}}_k = \gamma_2 K \sum \bar{a}_k \cdot \bar{\mathbf{v}}_k = \gamma_2 K \sum \overline{C}_k R_k \bar{\mathbf{v}}_k \tag{4.17}$$

with the shape factor $\overline{C}_k := \frac{C_k}{n_k \cdot \gamma_2 K}$ being independent of scale factors.

The purpose of this section is not to get a direct improvement of the quality of tidal solutions, but to understand the role of the normal modes in the composition of tidal oscillations and to analyze the effect of the LSA. The synthesized M_2-tide (Figure 4.1) can be compared with the solution of tidal models, e.g. that of [62] who utilized an equivalent discretization, the same boundary conditions and frictional parameterization and no data assimilation but the parameterized LSA-effect (see his Figure 1). There is a very good accordance between these two tidal solutions. All amphidromies are nearly at the same positions and also the amplitudes are quite similar. Thus, it is warranted that the M_2-tide can be sufficiently described by the superposition of the normal modes in the period range from 9 to 40 hours. The same applies to the other tidal constituents (not shown here).

4.1.2 LSA-effect on Forced Oscillations

The tidal syntheses are performed by using free oscillations computed with and without the LSA-effect . The expansion coefficients $\bar{a}_k$ and their components, the resonance depths R_k and the shape factors $\bar{C}_k$, of the synthesized tides are analyzed below.

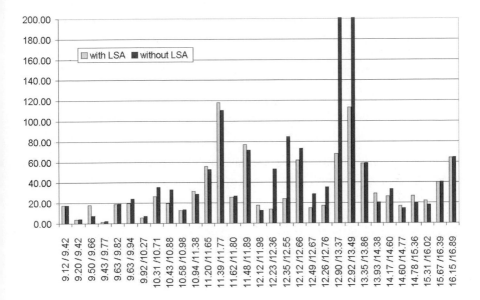

Fig. 4.2 The shape factors $\bar{C}_k$ of individual normal modes, when forced by the second degree semidiurnal potential. Black bars show these values for the modes calculated with LSA, gray bars for those calculated without LSA. The abscissa is marked by the periods of the normal modes calculated without/with LSA. The shape factors of the 12.90- and 12.92-mode are 468 and 518, respectively. The bars end at 200, to save space.

The LSA-effect in general

A general effect of LSA on resonantly forced oscillations is anticipated in this section. A theoretical approach [23] shows a decrease of the oscillationary part of the eigenfrequency σ_2 of the free oscillations through the LSA-effect . Utilizing now the approximation of the LSA-effect through $-0.085g\nabla\zeta$ (Accad and Pekeris [1978]) or replacing g in the Laplace tidal equations with $0.915g$. This leads to an effective reduction in g and can quantitatively explain the decrease of the frequencies of the gravitational modes, since the frequency is proportional to $\sqrt{g}$ for gravitational modes in idealized square basins.

Consequently, the resonance depth R_k is changed. Since the resonance depth is a complex value, both the intensity and the phase of the excited oscillation is influenced. The changes in the intensity were already discussed by [60]. He used an analytical hemispheric ocean model with the result that 'for individual near-resonance tidal constituents the rate of tidal power, e.g., can be reduced or enhanced by more than a factor two'. For the tidal oscillations of the World Ocean it will be discussed in more detail in the following subsections.

An already known, but so far not explained effect of the LSA on tidal dynamics, is that the tides are generally getting delayed [1, 56] when allowing for LSA. The explanation for this property is the LSA induced frequency shift, as well. This can be realized by writing the complex resonance depth as $R_k = \overline{R}_k e^{i\varphi_k}$; then $\overline{R}_k$ depicts the intensity and φ_k the phase given through R_k. Approximately, the phase shift induced by the LSA can be written as (Appendix 5):

$$\delta\varphi_k = \arctan\left(\frac{\sigma_{k,2}^{(noLSA)} - \sigma_{k,2}^{(LSA)}}{\sigma_{k,1}\left(1 + (\sigma_{k,2}^{(noLSA)} - \sigma_F)(\sigma_{k,2}^{(LSA)} - \sigma_F)/\sigma_{k,1}^2\right)}\right) \tag{4.18}$$

The delay gets large if both modes, with and without LSA, have frequencies close to the forcing frequency $((\sigma_{k,2}^{(noLSA)} - \sigma_F)(\sigma_{k,2}^{(LSA)} - \sigma_F)$ is small) and the frequency shift $\sigma_{k,2}^{(noLSA)} - \sigma_{k,2}^{(LSA)}$ induced by the LSA is large. This frequency shift is positive for all gravitational modes of the World Ocean since the periods of all these modes are lengthened through the LSA (Chapter 3). Thus, the phase shift is positive as well. For some near-resonant modes the phase delay amounts up to 92 and 65 degrees for the semidiurnal and diurnal tides, respectively (Figures 4.3 and 4.7).

Semidiurnal Tides

The potential of the semidiurnal tides is given by (4.13). The shape factors of each normal mode are equal for all semidiurnal tides of second degree, since they are independent of frequencies and of scale factors. The effect of LSA on the shape factors is quite large for six of the normal modes in the semidiurnal spectrum. These six modes have periods of 12.36, 12.55, 12.67, 12.76, 13.37, and 13.49 hours (Figure 4.2). The effect of LSA is largest for the 13.37- and 13.49-mode (Figure 5.8). There, the shape factors are reduced through LSA by a factor 6.9 and 4.6, respectively. These two modes are similar in their spatial patterns of ζ and $\mathbf{u}$. In both cases, whether LSA is considered or neglected, the two modes are forced in opposite phase and consequently weaken each other. Although the 13.37- and 13.49-mode are much less excited than the corresponding pair computed without LSA (12.90- and 12.92-mode) the resulting effect is small, since they diminish their combined contribution. Quantitatively, we can express this with an 'effective' expansion coefficient $\overline{a}_{eff}$. This coefficient is defined for the k-th and the l-th mode as $\overline{a}_k \mathbf{v}_k + \overline{a}_l \mathbf{v}_l = \overline{a}_{eff} \mathbf{v}_{eff}$, with the constraint $\sqrt{\int_S \zeta_{eff} \cdot \zeta_{eff}^* dA} = 1$ (see also (4.16)). The ratio $\frac{|\overline{a}_{eff}|}{|\overline{a}_k| + |\overline{a}_l|}$ is a measure for the strength of the interference (1=full constructive interference, 0=full

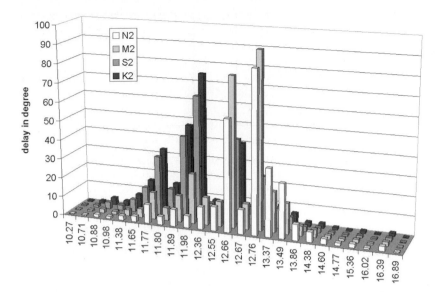

Fig. 4.3 The phase delays $\delta\varphi_k$ of individual normal modes induced by the LSA-effect when forced by the semidiurnal potentials of second degree (K_2, S_2, M_2, N_2).

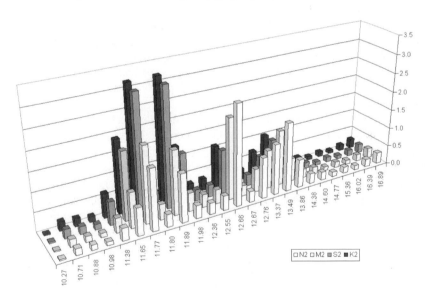

Fig. 4.4 The expansion coefficients $\overline{a}_k$ of individual normal modes when forced by the semidiurnal potentials of second degree (K_2, S_2, M_2, N_2).

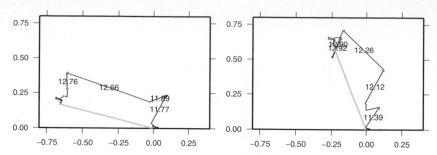

Fig. 4.5 The composition of the M_2 tidal wave along the African coast analyzed at the point $20°S/10°E$; **left**: with full LSA; **right**: neglecting the LSA; each arrow represents one normal mode. The length of the arrow is equivalent to the amplitude of the sea surface elevation ζ. The direction represents the phase of ζ. Time goes in the clockwise direction. The arrows have arrow heads and are marked by the respective periods of the normal modes when the amplitude is larger than 10cm. The arrow in gray represents the resulting tidal sea surface elevation. The scale of the axes is in [cm].

destructive interference). In case of the M_2 constituent and the above mentioned mode pair computed with and without LSA, it takes the values 0.39 and 0.07, respectively. This reflects the global destructive interference, especially for the two modes calculated without LSA.

Having in mind the prolongation of the periods through the LSA-effect of around δT=0.5 hours, it is clear that the magnitude of the resonance depth R_k is enhanced for modes with periods lower than the forcing period T_F and reduced for modes with period larger than $T_F + \delta T$. The change of the resonance depths of the normal modes with periods in between $[T_F, T_F + \delta T]$ depends on the individual values of these modes.

The influence of LSA on the phases of the resonance depths has been generally discussed in the beginning of this section. In Figure 4.3 the phase delays due to LSA are shown in case of semidiurnal forcing. The four normal modes 11.89, 11.97, 12.66, and 12.76 are delayed by more than 40 degrees. In case of M_2 the 12.66- and 12.76-mode (Figures 5.6 and 5.7), are forced with very large phase delays of 78 and 92 degrees, respectively. In the synthesis of the M_2-tide the 12.66-mode plays a major role (Figure 4.4). The ascertained phase delays of M_2 in tidal models when considering LSA (e.g. [1, 56]) are mainly ascribed to the large phase delay of the 12.66-mode. Analysis of [57] showed phase delays in the M_2 pattern along the African coast of up to 70 degrees. In the present study the composition of this tidal wave is analyzed at $(20°S/10°E)$ (Figure 4.5). It exemplarily shows the influence of the delays of the 12.66-mode and the 12.76-mode on the composition of the M_2-tide. At this location, close to the African Coast, the synthesized M_2-tide is delayed by around 54 degrees. Furthermore, Figure 4.5 makes obvious the destructive interference of the 12.90- and 12.92-mode, discussed in the beginning of this section.

Diurnal Tides

The shape factors determined by the use of the second degree diurnal forcing are shown in Figure 4.6. Here, the influence of LSA is small, contrary to that on on the shape factors of the semidiurnal forcing. Only two modes (27.19- and 28.20-mode, Figure 5.14 and 5.15) show a large influence of LSA on their shape factors. The major influence of LSA is on the resonance depth . Especially the resonance depth of the 26.20-mode differs from that of the corresponding mode determined without LSA (24.61-mode). All of the other near-resonant modes have smaller frequency shifts (smaller than 1 hour), resulting in a smaller influence of the LSA on their resonance depths.

The LSA induced phase delay is smaller for the diurnal tides than for the semidiurnal tides due to the following reasons. Firstly, the frequency of one of the dominant modes, the 32.64-mode, is not close to the forcing frequencies (Figure 4.7). Secondly, the 23.95-mode and 27.18-mode are topographical vorticity modes and are negligible in the construction of the diurnal tides, due to their small decay times and small shape factors (4.8). Anyway, they have very small decay times resulting in a small LSA induced phase delay (4.18). The 25.32 and 27.57-mode have also small decay times which lead hereby to reduced resonance depths (4.8) and phase delays (4.18). The remaining 26.20-mode (Figure 5.13) is the only mode, which has phase

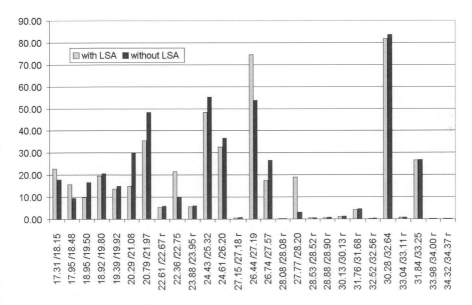

Fig. 4.6 The shape factors $\overline{C}_k$ of individual normal modes, when forced by the second degree diurnal potential. Black bars show these values for the modes calculated with LSA, gray bars for those calculated without LSA. The abscissa is marked by the periods of the normal modes calculated without/with LSA, the appended "r" indicates topographical vorticity modes.

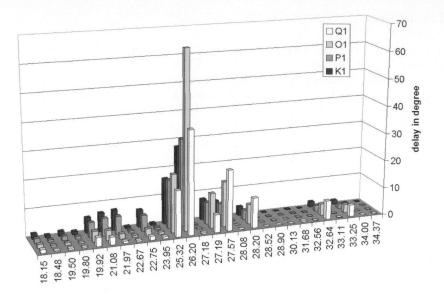

Fig. 4.7 The phase delays $\delta\varphi_k$ of individual normal modes induced by the LSA-effect when forced by the diurnal potentials of second degree (K_1, P_1, O_1, Q_1).

delays larger than 25 degrees. The largest appears in the case of the O_1 forcing, namely 65 degree. For this tide the 26.20-mode has its strongest resonance.

4.1.3 The Synthesis of the Semidiurnal and Diurnal Tides

Semidiurnal Tides

In the semidiurnal band 1.75-2.25 cpd (10.7-13.7h), as defined by PL1981, are 16 modes. Nine of them have small shape factors and are playing a minor role.

The expansion coefficients of the synthesized semidiurnal tides of second degree are shown in Figure 4.4. The 12.66-mode (Figure 5.6) is the most strongly excited in case of the M_2 and N_2 tidal forcing due to its large resonance depth and shape factor (Figure 4.2). The 11.77- and the 11.89-mode (Figure 5.4 and 5.5) have large shape factors and consequently, are important for M_2 and N_2, although they have lower resonance depths. These two modes have large resonance depths in case of the S_2 and K_2 forcing. Thus they are dominating these two tidal constituents. All three modes mentioned so far have the largest part of their energy in the Atlantic Ocean. The 12.66-mode has 32% of its energy in the Pacific. There, it appears as an equatorial standing wave of 3/2 wavelength.

In the Atlantic Ocean, strong local resonances for the semidiurnal tides are well known, e.g. the resonances in the Gulf of Maine or in the Amazon estuary. These

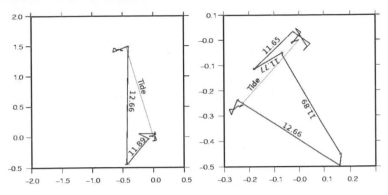

Fig. 4.8 The composition of the semidiurnal resonance in the Gulf of Maine ($45°N/74°W$); **left**: M_2; **right**: S_2 forcing. Information given as in Figure 4.5, except for the left and right figure the arrows marked by the respective periods of the normal modes when the amplitude is larger than 30cm and 10cm, respectively.

resonances appear in the patterns of some free oscillations as well. [11] estimated the period of the free oscillation of the Bay of Fundy - Gulf of Maine system to 13 hours. He assumed that the semidiurnal response of the system is largely determined by one mode. Due to the one degree resolution of the model in the present study, the Gulf of Maine is only roughly resolved. Nevertheless, strong resonances appear in a few normal modes within this region. The dominant mode at this location is, in case of M_2, the 12.66-mode and a small amount comes from the 11.89-mode. For S_2, the 11.89-mode gains influence and lies level with the 12.66-mode (Figure 4.8). Another strong resonance appears in the equatorial region of the Atlantic Ocean, namely in the Amazon estuary. There, the resonance is predominantly due to the 12.66-mode as well. The two other prominent contributors, 11.77- and 11.89-mode, to a great extent cancel out.

In the Pacific, the New Zealand Kelvin Wave is a prominent feature of the semidiurnal tides. This wave appears in a lot of modes in the semidiurnal period range. In the 11.98-mode, this feature is most pronounced. In the style of PL1981 (mode 38, 10.8h ,cf. Fig 20) it can be named as the free New Zealand Kelvin Wave. However, both the 11.65- and the 11.89-mode largely contribute to this Kelvin Wave.

In the Indian Ocean a subtropical midocean antinode is predicted for all of the four main semidiurnal constituents. In the synthesized tide of [35] this feature appears as well, but it could not be assigned to individual normal modes. The present study shows clearly that the Indian antinode appears in the normal modes with frequencies lower than 11.7h (e.g. 11.65h, 11.38h, and 10.98h) and the main contribution comes from the 11.65-mode (Figure 5.3).

Diurnal Tides

In [35] there are only five modes in the diurnal band 0.75-1.25 cpd (19.2-32.0h). Two of them are suppressed by their small shape factors ('cross sections') for the second-degree diurnal potential. Thus, 'the spectral composition of diurnal tides is relatively pure' [35]. In the present study 11 gravitational modes are found in the diurnal band, plus the 32.64-mode being strongly excited although its frequency is not close to the forcing frequeny. A good representation of the diurnal tides is derived by synthesizing these modes. In particular an accurate spectral composition of the oscillation system of the Indian Ocean, the AKW1 and the topographical

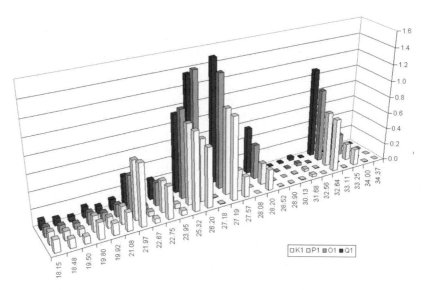

Fig. 4.9 The expansion coefficients $\bar{a}_k$ of individual normal modes when forced by the diurnal potentials of second degree (K_1, P_1, O_1, Q_1).

trapping of a vorticity mode at the New Zealand Plateau results.

The relative importance of individual normal modes of the synthesized diurnal tides is quantitatively expressed by their expansion coefficients (4.8) and is shown in Figure 4.9. A conspicuous feature is that the 32.64-mode (Figure 5.15) with its first order Antarctic Kelvin Wave (AKW1) plays an important role in the construction of the diurnal tides of second degree, although the frequency of this mode is not close to the forcing frequencies. The strong excitation of this mode is due to the large shape factor (Figure 4.6). The AKW1 is a significant feature of all diurnal tides of second degree. For the O_1-tide the AKW1 is analyzed at two positions ($50°S/50°E$ and $50°S/0°E$) located in the Southern Ocean. The extent of contribution of individual normal modes to the sea surface elevation ζ at these locations is shown in Figure 4.10. The main contributors are the 26.20-mode and the 32.64-mode (Fig-

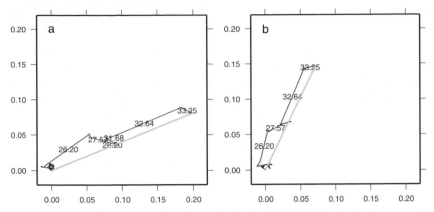

Fig. 4.10 The spectral composition of the first order Antarctic Kelvin Wave in the O_1 pattern; left: at the position $50°S/50°E$; right: at the position $50°S/0°E$. Information given as in Figure 4.5, except for the arrows marked by the respective periods of the normal modes when the amplitude is larger than 1.5cm.

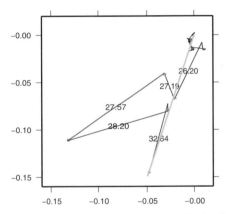

Fig. 4.11 The spectral composition of the resonance at the New Zealand Plateau in the O_1 pattern at the postion $45°S/10°W$. Information given as in Figure 4.5, except for the arrows are marked by the respective periods of the normal modes when the amplitude is larger than 2.5cm.

ures 5.13 and 5.15). Both modes are nearly in phase in that region, resulting in a constructive interference. The 26.20-mode has only 4% of its total energy in the Southern Ocean. It is strongly excited, especially in case of $O1$ and $Q1$, due to its closeness to the forcing periods. According to this, it is an important contributor to the AKW1, although the AKW1 is most pronounced in modes with periods longer than 32 hours (Section 3.1.3). In [35], the mode 16 (28.7h, cf. Fig.17, Pl1981) contributes almost exclusively to the AKW1. This mode contributes also the largest portion to that synthesis of O_1 and K_1. It is most likely resembled by the 32.64-mode of the present study.

The modes with periods of 25.32h and 27.19h (Figure 5.13 and 5.14) are fundamental Pacific modes (Section 3.1.2). They are not captured by the models of

ZAMU2005 and PL1981 and thus, are not included in their tidal analysis. Both are playing an important role in the formation of the diurnal tides in the North Pacific. There, they appear as a Kelvin Wave along the North Pacific Coast. Moreover, the 21.97-, 26.20-, 27.57-, 28.20-, and 32.64-mode contribute to this Kelvin Wave, where the 27.57- and the 28.20-mode weaken the amplitude. In the North Pacific, the latter ones are nearly in phase, but in the South Pacific, namely at the New Zealand Plateau, they are in opposite phase (Figure 4.11). There, these two modes are affected by a topographical vorticity mode (Section 3.1.2). Obviously, they mutually diminish their contribution to the vorticity mode at the New Zealand Plateau. However, the topographical vorticity mode keeps preserved in the patterns of the diurnal tides, since the AKW1 (via the 32.64-mode) is also affected by this vorticity mode (see Section 3.1.2)

In the Indian Ocean and the adjacent Southern Ocean a single amphidrome appears in the tidal patterns of O_1 and Q_1, located at approximately $30°S$. In case of K_1 and P_1 this amphidrome is shifted to $50°S$, and an additional amphidrome appears at the equator. All free oscillations with periods longer than 23 hours have a single amphidromic point in the Indian Ocean. The 19.80-mode (Figure 5.11) has three amphidromic points instead. Thus, this mode is obviously the main reason for the amphidromic point near south of India, which appears in the K_1- and P_1-tide. This hypothesis is corroborated by the fact that the second amphidrome disappears in the synthesis of the K_1-tide when neglecting the 19.80-mode. Interesting to note is that tidal ocean models with assimilation of satellite data predict two amphidromic points in the patterns of O_1 in the Indian Ocean (e.g. [62]), as well.

4.2 Integration of the Solutions of a Tidal Model with Assimilation of Data

The assimilation of tidal sea surface elevations extracted from satellite altimetry (TOPEX/POSEIDON) provided means to improve the solutions of tidal ocean models and analyze the failure of free tidal models in terms of the features of dynamical residues (e.g. [63, 6]).

So far, a method to improve the free barotropic oscillations of the World Ocean through assimilation of data is not known. The main problem is that for periodically forced 'free' oscillations no measurements are available. In case of a periodical forcing the free oscillations adapt the frequency of the forcing. Thus, the forced oscillation results as a composition of inseparable free oscillations. In case of stochastically forced free oscillations a few measurements of single free oscillations are available, since the eigenfrequency keeps preserved [28].

In this section the focus is on the gravitational free oscillations forced by the semidiurnal and diurnal tidal potential. [11] presented a method to determine the period and the damping rate of a tidally forced (near-resonant) normal mode in the Bay of Fundy/Gulf of Maine system, by using records of sea level of a few semidiurnal tides at different locations. A theoretical discussion of this approach is given by [52]. Similarly, [42] estimated periods of free oscillations responsible for the resonant third-degree diurnal tides in the North Atlantic. Recently, the method of [11] was further developed by [49] in order to determine the frequency and the damping rate of a non-resonant forced free oscillation in the Juan de Fuca Strait. All these approaches have in common that the response of the system is largely in one mode or at most in two modes.

The purpose of the present study is to combine the results of a tidal model assimilating satellite data [63], hereinafter referred to as ZA2000) with that of an ocean model determining the barotropic free oscillations of the World Ocean (Chapter 3). For this combination the synthesis procedure derived in Section 4.1 is utilized and two distinct methods are developed to extract information out of the results of the tidal model. On the one hand a method is derived to directly obtain improved expansion coefficients for the synthesis of tides. On the other hand the tidal solutions are used for nonlinear least squares fits in order to obtain more realistic estimates of the eigenfrequencies of some free oscillations in the semidiurnal and diurnal period range. An accessory parts of the latter method is an adjustment of the corresponding adjoint eigenfunctions.

4.2.1 New Expansion Coefficients

Every periodically forced oscillation $\mathbf{w}_F$ can be expressed through a superposition of free oscillations $\mathbf{v}_k$ (Section 4.1, equations 4.3 and 4.8):

$$\mathbf{w}_F = \sum_{k=1}^{\infty} a_k \cdot \mathbf{v}_k \tag{4.19}$$

$$a_k = \frac{1}{i(\sigma_k - \sigma_F)} \cdot \langle \mathbf{F}, \widehat{\mathbf{v}}_k \rangle \tag{4.20}$$

with the expansion coefficients a_k and the eigenfrequency σ_k of the k-th free oscillation $\mathbf{v}_k$, the frequency σ_F of the external forcing and the corresponding adjoint eigenfunction $\widehat{\mathbf{v}}_k$.

Assuming now that the forced oscillation is known, e.g. the solution $\mathbf{w}_F^{(a)}$ of a tidal model with assimilation of data. Multiplying (4.19) scalarly with $\widehat{\mathbf{v}}_l$ and using the orthogonality relation (4.6) leads to the estimates of the expansion coefficients

$$a_l^{(a)} = \langle \mathbf{w}_F^{(a)}, \widehat{\mathbf{v}}_l \rangle. \tag{4.21}$$

In doing so, the main assumption is that the free oscillations of the oscillation system described through $\mathbf{w}_F^{(a)}$, can be approximated to the first order by those of the barotropic ocean model. The tidal solutions $\mathbf{w}_F^{(a)}$ used in this study are from ZA2000. Their bathymetry and the one degree finite difference discretization is exactly the same, which is used for computing the biorthogonal system of eigenfunctions used in the present study. Since the solutions of tidal models with assimilation of data are differng only in some details [44], it is expected that the following analyses are independent of the choice of tidal model solutions.

For the above mentioned estimation of the expansion coefficients, four semidiurnal (K_2, S_2, M_2, N_2) and four diurnal (K_1, P_1, O_1, Q_1) tidal constituents are available. For all these constituents the estimates of the expansion coefficient $a_k^{(a)}$ are computed and are shown in Figures 4.15 and 4.22. Further, the syntheses of the tides are performed by means of the 'new' expansion coefficients $a_k^{(a)}$

$$\tilde{\mathbf{w}}_F^{(a)} = \sum a_k^{(a)} \cdot \mathbf{v}_k. \tag{4.22}$$

Synthesis of tides

In Figure 4.12, the tidal patterns of the M_2-tide of the new synthesized M_2-tide $\tilde{\mathbf{w}}_F^{(a)}$, of the synthesized M_2-tide $\tilde{\mathbf{w}}_F$ of Section 4.1 and the one obtained by ZA2000 are shown.

Generally speaking, the assimilation of data compensates the overestimation of the amplitudes of the tides in free tidal models (e.g. [62]). This correction appears in the synthesized M_2-tide, as well, when using the new expansion coefficients $a_k^{(a)}$. It appears mainly as a reduction of the amplitudes, e.g. of those of the equatorial Pacific 3/2 transverse half-wave or those of the whole Atlantic. Further, the corrections of the amphidromic structure in the North Pacific, brought about by the assimilation of data, is well displayed through the new synthesized tide. However, the disappearance of the amphidrome south of Australia is not caught by this synthesis, but is

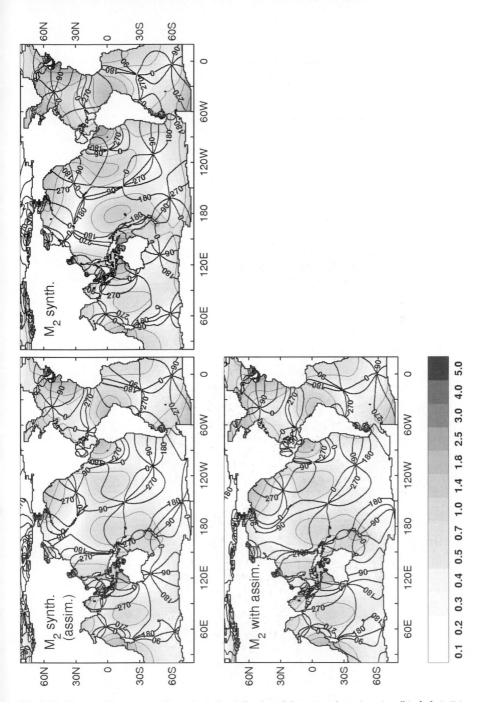

Fig. 4.12 The M_2-tide patterns. Gray shaded amplitudes of the sea surface elevation ζ in $[m]$. Solid lines showing the phases in degrees (in steps of $30°$). **Upper left**: New synthesized M_2-tide $\tilde{\mathbf{w}}_F^{(a)}$; **upper right**: Synthesized M_2-tide $\tilde{\mathbf{w}}_F$ of Section 4.1; **lower left**: the M_2-tide of the tidal model of ZA2000.

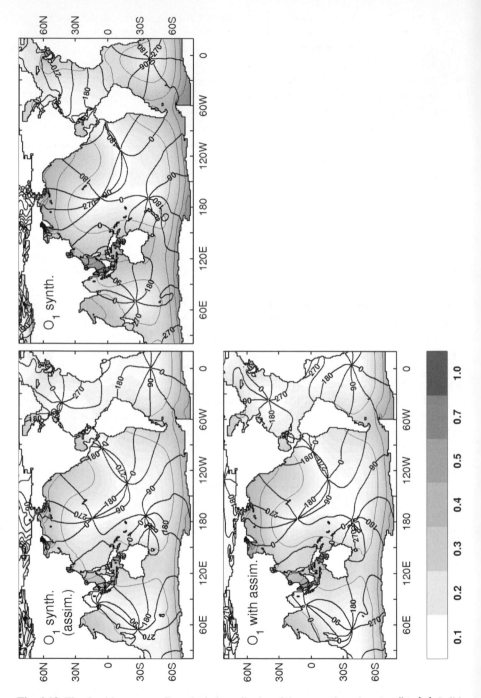

Fig. 4.13 The O_1-tide patterns. Gray shaded amplitudes of the sea surface elevation ζ in $[m]$. Solid lines showing the phases in degrees (in steps of $30°$). **Upper left**: New synthesized O_1-tide $\tilde{\mathbf{w}}_F^{(a)}$; **upper right**: Synthesized O_1-tide $\tilde{\mathbf{w}}_F$ of Section 4.1; **lower left**: the O_1-tide of the tidal model of ZA2000.

obviously initiated. The existence of this amphidrome as a fully developed one is attributed to the roughly resolved bathymetry of the one degree ocean model. [54] showed that the disappearance of the amphidrome south of Australia is very sensitive to the choice of the bathymetry and its resolution. Following this line of thoughts yields that this amphidrome would disappear in the patterns of the relevant free oscillations, as well, when refining the gridsize and using an appropriate bathymetry. Consequently, in this case it would also disappear in the patterns of the synthesized M_2-tide. The disappearance of the amphidrome in the Southern Ocean north of the Ross Sea, as a consequence of assimilating data, is not sufficiently represented by the synthesized tide, as well. However, it is shifted correctly in southward direction towards Antarctica, when using the new expansion coefficients.

As a representative of the diurnal tides, the tidal patterns of the O_1, are shown in Figure 4.13, as obtained by the tidal model of ZA2000 and through the two synthesizing methods (of Section 4.1 and that of the present Section). The amplitude reduction of the Antarctic Kelvin Wave through assimilation of data is well represented in the new synthesized tide, as well as the appearance of an amphidrome south of India. Further, the assimilation of data brought about a shift of the South Indian amphidrome in south-west direction. This shift appears as well in the new synthesized O_1, but it is not large enough. In the equatorial Pacific the developing of a northward propagating wave along the American coast appears in the synthesized tide when using the newly obtained expansion coefficients, consistent with the result of ZA2000. The amphidromic system of the Atlantic north of the equator comes into existence but is not fully marked in the synthesized tide compared to the results of the tidal model. In the South Atlantic, the cyclonically rotating single amphidrome is delayed by around 45 degrees and slightly shifted eastward resulting in a north-easterly propagating wave in the equatorial region, according to the tidal model of ZA2000.

4.2.2 New Frequencies and Adjoint Eigenfunctions

An estimate of the frequency of the k-th free oscillation and of the adjoint eigenfunction $\widehat{\mathbf{v}}_k$, can be obtained by minimizing the difference between (4.20) and (4.21), i.e. the function

$$Q(\sigma_k^{(a)}, \widehat{\mathbf{v}}_k^{(a)}) = \sum_{pt=1}^{8} \left(\frac{1}{i(\sigma_k^{(a)} - \sigma_{F,pt})} \cdot \langle \mathbf{F}_{pt}, \widehat{\mathbf{v}}_k^{(a)} \rangle - \langle \mathbf{w}_{F,pt}^{(a)}, \widehat{\mathbf{v}}_k^{(a)} \rangle \right)^2$$
$$+ \sum_{i=1}^{N} \left(\widehat{v}_{k,i}^{(a)} - \widehat{v}_{k,i} \right)^2 + \left(\sigma_k^{(a)} - \sigma_k \right)^2. \quad (4.23)$$

There, $\widehat{\mathbf{v}}_k^{(a)}$ and $\sigma_k^{(a)}$ are the estimates of the k-th adjoint eigenfunction $\widehat{\mathbf{v}}_k$ and frequency σ_k, respectively, and $pt = 1, 2, ..., 8$ stands for the eight partial tides, which are available for this analysis. The two latter terms of equation (4.23) are appended in order to keep the estimates close to the original solutions of $\widehat{\mathbf{v}}_k$ and σ_k. This is a nonlinear least squares problem with N+1 complex unknowns. In the present study the nonlinear least squares problem is divided into two subproblems: First, a nonlinear one with one complex unknown, namely the frequency $\sigma_k^{(a)}$:

$$Q_1(\sigma_k^{(a)}, \varepsilon) = \sum_{pt=1}^{8} \left(\frac{1}{i(\sigma_k^{(a)} - \sigma_{F,pt})} \cdot \langle \mathbf{F}_{pt}, \widehat{\mathbf{v}}_k^{\varepsilon} \rangle - \langle \mathbf{w}_{F,pt}^{(a)}, \widehat{\mathbf{v}}_k^{\varepsilon} \rangle \right)^2$$
$$+ \left(\sigma_k^{(a)} - \sigma_k \right)^2 \qquad (4.24)$$

and a linear one to determine the N complex unknowns $\widehat{\mathbf{v}}_k^{(a)}$

$$Q_2(\widehat{\mathbf{v}}_k^{(a)}, \varepsilon) = \sum_{pt=1}^{8} \left(\frac{1}{i(\sigma_k^{\varepsilon} - \sigma_{F,pt})} \cdot \langle \mathbf{F}_{pt}, \widehat{\mathbf{v}}_k^{(a)} \rangle - \langle \mathbf{w}_{F,pt}^{(a)}, \widehat{\mathbf{v}}_k^{(a)} \rangle \right)^2$$
$$+ \sum_{i=1}^{N} \left(\widehat{v}_{k,i}^{(a)} - \widehat{v}_{k,i} \right)^2, \qquad (4.25)$$

there $\varepsilon = 1, 2, ...$ denotes the iteration step. The algorithm is shown in Figure 4.14. To solve the linear problem (4.25), the Conjugate Gradient Least Squares method

Input $(\sigma_k, \widehat{\mathbf{v}}_k)$
Put $\sigma_k^1 = \sigma_k$
For $j = 1, 2, 3, ...$

    (1) Minimizing: $Q_2(\widehat{\mathbf{v}}_k^j, \varepsilon = j)$
    (2) Minimizing: $Q_1(\sigma_k^j, \varepsilon = j)$
    (3) $\sigma_k^{j+1} = \sigma_k^j, \widehat{\mathbf{v}}_k^{j+1} = \widehat{\mathbf{v}}_k^j$
End For

Fig. 4.14 Algorithm: Frequency and adjoint eigenfunction estimation.

(CGLS) described in [32] is used. The nonlinear problem (4.24) is solved with the Levenberg Marquardt Algorithm [25]. This analysis is only suitable for near-resonant free oscillations with large expansion coefficients. The minimization of (4.24) and (4.25) is performed for selected modes in the semidiurnal and in the diurnal period ranges. The results of the frequencies $\sigma_k^{(a)}$ are shown in Table 4.1 and 4.2. The residues E, defined through

$$E_k = \sum_{pt=1}^{8} \left(\frac{1}{i(\sigma_k^{(a)} - \sigma_{F,pt})} \cdot \langle \mathbf{F}_{pt}, \widehat{\mathbf{v}}_k^{(a)} \rangle - \langle \mathbf{w}_{F,pt}^{(a)}, \widehat{\mathbf{v}}_k^{(a)} \rangle \right)^2 \qquad (4.26)$$

are given in the last column. Further, the associated resonance curves given through

$$A_{\sigma_k}(\sigma) = \frac{1}{i(\sigma_k - \sigma)}, \tag{4.27}$$

are plotted for a few modes, with σ_k being the original eigenfrequency and $\sigma_k^{(a)}$ the new one. Additionally, in these figures the values $P^{(a)}(\sigma_{F,pt}) = \frac{\langle \mathbf{w}_{F,pt}^{(a)}, \widehat{\mathbf{v}}_k^{(a)} \rangle}{\langle \mathbf{F}_{pt}, \widehat{\mathbf{v}}_k^{(a)} \rangle}$ and $P(\sigma_{F,pt}) = \frac{\langle \mathbf{w}_{F,pt}^{(a)}, \widehat{\mathbf{v}}_k \rangle}{\langle \mathbf{F}_{pt}, \widehat{\mathbf{v}}_k \rangle}$ are marked. $P(\sigma_{F,pt})$ illustrates the deviation of the expansion coefficients obtained through (4.21) from the original values (4.20) and $P^{(a)}(\sigma_{F,pt})$ shows these values after minimizing the function (4.23).

4.2.3 Results

Semidiurnal Tides

This new approach of synthesizing tides yields in case of the two semidiurnal forcings of K_2 and S_2 that almost all expansion coefficients are too large, which means that their contributions are overestimated. In case of the tidal constituents M_2 and N_2 the modes with periods greater than 12.5 hours have too large expansion coefficients, as well. In Figure 4.15 the absolute values of the expansion coefficients of all modes in the semidiurnal period range are given. The difference between the estimated $\bar{a}_k^{(a)}$ and the expansion coefficients $\bar{a}_k$ of Section 4.1 are shown in Figure 4.16. Further, least squares fits as described in the previous section, have been performed for selected modes (Table 4.1).

The expansion coefficients of the 11.65-mode are nearly unchanged for M_2 and N_2 and somewhat reduced for K_2 and S_2. However, the relative influence of this mode is enhanced, especially for the K_2- and S_2-tides, due to the fact that the two dominant modes (11.77- and 11.89-mode, see below) are by far stronger weakened. The least squares fit yields a somewhat reduced period of 11.4 h and a nearly halved decay time, amounting to 37 hours. The estimated ζ-values of the adjoint mode $\widehat{\mathbf{v}}_k^{(a)}$ are shown in Figure 4.17 and it can be recognized that the most significant changes occur in the Pacific and in the adjacent Southern Ocean.

The influence of the 11.77- and 11.89-mode, the two dominant modes for the K_2- and S_2-tides, is strongly reduced in the solutions of ZA2000. For M_2 the influence keeps nearly preserved and for N_2 it is enhanced. Figure 4.18 shows the resonance curve of the 11.77-mode. The resonance curve $A_{\sigma_k^{(a)}}(\sigma)$ is flattened due to the reduction of the damping rate from 26 hours to 13 hours and the period is nearly unchanged.

The 12.36-mode gains influence for the two slower tidal constituents, and for S_2 and K_2 the influence remains unchanged. The least squares fit yields an enlarged period of around 12.8 hours and the decay time is roughly doubled to around 48 hours. The resonance curve (4.19) shows that the least squares fit predominantly changes

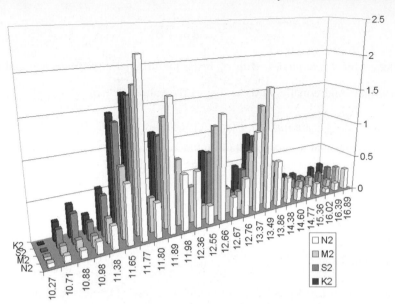

Fig. 4.15 The expansion coefficients $\bar{a}_k^{(a)}$ of individual normal modes, when forced by the semidiurnal potentials of second degree (K_2, S_2, M_2, N_2).

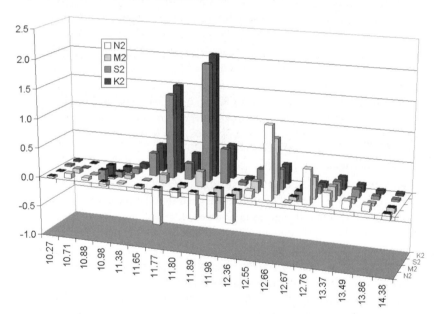

Fig. 4.16 The difference $\bar{a}_k - \bar{a}_k^{(a)}$ between the expansion coefficients $\bar{a}_k$ of the tidal synthesis of Section 4.1 and the coefficients $\bar{a}_k^{(a)}$, when forced by the semidiurnal potentials of second degree (K_2, S_2, M_2, N_2).

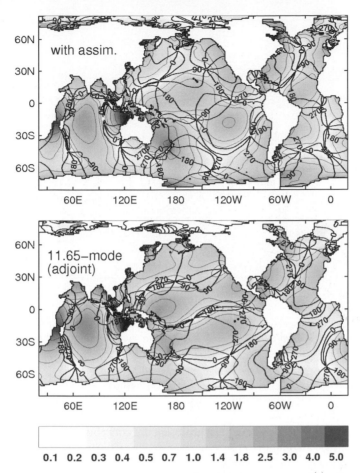

Fig. 4.17 The adjoint 11.65-mode (**bottom**) and the corresponding mode $\widehat{\mathbf{v}}_k^{(a)}$ (**top**) determined through minimizing (4.24) and (4.25). Gray shaded amplitudes of the sea surface elevation ζ in [m]. Solid lines showing the phases in degrees (in steps of 30°).

the frequency and not the adjoint mode, since the P and $P^{(a)}$ values are lying close together.

The 12.66-mode looses its dominant role for the M_2- and N_2-tides and its influence is reduced for all four semidiurnal constituents. The least squares fit changes the frequency to fit the amplitude through the given values of $|P|$, and the adjoint mode is adjusted to approximate the $P^{(a)}$ to the resonance curve (Figure 4.20). The period is changed to 12.9 hours and the decay time is reduced from 38 to 22 hours, which accounts for the reduced influence for the semidiurnal tides of this mode. The adjoint mode resulting through the least squares fit is shown in Figure 4.21. An interesting detail is that the amphidrome south of Australia disappears. This amphidrome disappears in the solutions of tidal models with assimilation of data, as well. Of course,

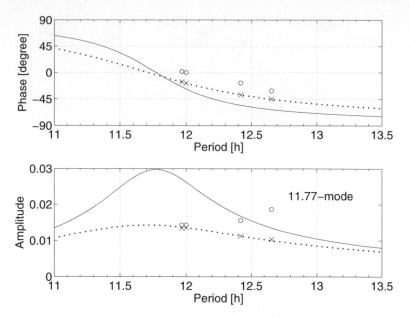

Fig. 4.18 The resonance curve $A_{\sigma_k}(\sigma)$ of the 11.77-mode according to equation (4.27). **Top**: phase τ defined through $tan(\tau) = Im(A(\sigma))/Re(A(\sigma))$. **Bottom**: amplitude defined through $|A(\sigma)|$. Dotted lines represent the resonance curve $A_{\sigma_k^{(a)}}(\sigma)$. Solid lines show the resonance curve with the original eigenfrequency σ_k. The x and o mark the values $P^{(a)}(\sigma_{F,pt})$ and $P(\sigma_{F,pt})$, respectively, for the semidiurnal tides (pt: K_2, S_2, M_2, N_2) (see text for explanation).

this is the adjoint mode, and does not represent the contribution of this mode to the ocean tides. But a comparison of the 12.66-mode with its corresponding adjoint mode shows many similarities, e.g. the amphidrome south of Australia. Therefore we might expect this amphidrome disappearing in the 12.66-mode as well.

The relative importance of the 13.37-mode and of the 13.49-mode is enhanced since their expansion coefficients remain nearly unchanged. A least squares fit for these two modes is senseless. Because of the small resonance depths and the small changes in the expansion coefficients, the fits would not be meaningful.

Table 4.1 Periods T_2 and damping rates T_1 obtained by nonlinear least squares fits for selected modes in the semidiurnal period range. Last column shows the residues (4.26).

T_2	$T_2^{(a)}$	T_1	$T_1^{(a)}$	E
11.65	11.43	62.68	36.80	1.05E-03
11.77	11.71	26.32	12.65	6.15E-04
11.98	11.65	68.38	37.65	2.05E-03
12.36	12.78	22.01	48.12	4.02E-03
12.66	12.88	37.60	21.60	1.25E-03

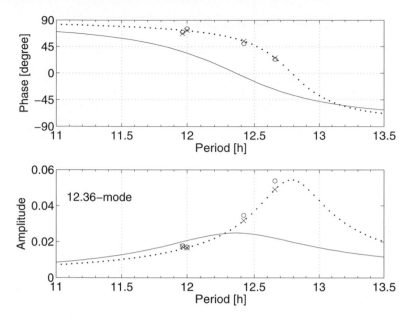

Fig. 4.19 The resonance curve $A_{\sigma_k}(\sigma)$ of the 12.36-mode according to equation (4.27). Information given as in Figure 4.18.

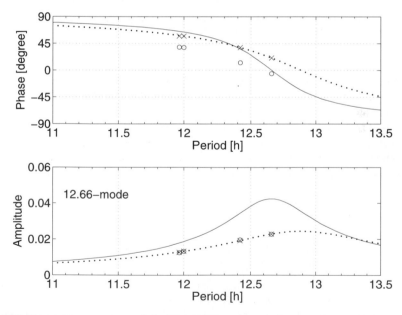

Fig. 4.20 The resonance curve $A_{\sigma_k}(\sigma)$ of the 12.66-mode according to equation (4.27). Information given as in Figure 4.18.

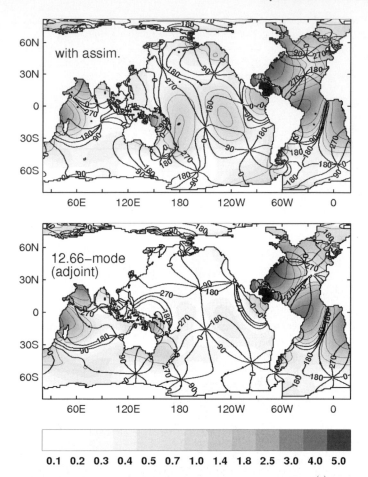

Fig. 4.21 The adjoint 12.66-mode (**bottom**) and the corresponding mode $\widehat{\mathbf{v}}_k^{(a)}$ (**top**) determined through minimizing (4.24) and (4.25). Gray shaded amplitudes of the sea surface elevation ζ in $[m]$. Solid lines showing the phases in degrees (in steps of $30°$).

Diurnal Tides

For all of the synthesized diurnal tidal constituents the contributions of modes with periods slower than 25 hours are overestimated and those of the faster ones are underestimated with respect to the synthesis obtained through integrating the results of ZA2000. These newly obtained expansion coefficients are shown in Figure 4.22 and the differences between the coefficients of Section 4.1 and those of the present section are shown in Figure 4.23. In Table 4.2 the results of the least squares fits of selected modes are given.

The 21.97-mode gains influence on the one hand through a larger expansion coefficient, and on the other hand through the fact that almost all other important modes

Table 4.2 Periods T_2 and damping rates T_1 obtained by nonlinear least squares fits for selected modes in the diurnal period range. Last row shows the residues (4.26).

T_2	$T_2^{(a)}$	T_1	$T_1^{(a)}$	E
21.97	23.67	36.47	22.42	1.80E-03
22.75	23.70	18.03	25.33	2.19E-03
25.32	29.36	19.79	18.17	7.01E-04
26.20	28.55	41.32	28.71	1.35E-03

are weakened. The least squares fit yields a period of 23.7 hours, thus it is shifted towards resonance. The decay time is reduced from originally 37 hours to 22 hours.

The expansion coefficients of the 22.75-mode are only slightly changed, instead

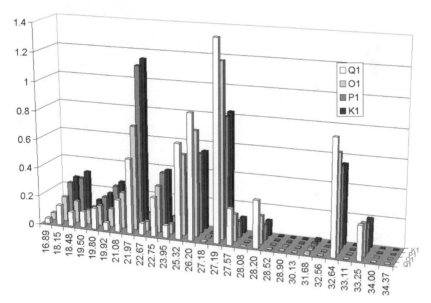

Fig. 4.22 The newly obtained expansion coefficients $\bar{a}_k^{(a)}$ of individual normal modes, when forced by the diurnal potentials of second degree (K_1, P_1, O_1, Q_1).

larger changes occur for the least squares fit of the frequency (Figure 4.26) and the adjoint eigenfunction (Figure 4.25). The fit yields a period of 23.7 hours and the somewhat enhanced decay time of 25 hours. For the adjoint eigenfunction signifi-cant changes occur in the Pacific and in the Southern Ocean. Both the 25.32-mode and the 26.20-mode are strongly overestimated in amplitude. For the 25.32-mode the least squares fit yields a strongly enlarged period of 29.4 hours and the decay time of around 20 hours keeps nearly unchanged. Instead the decay time of the 26.20-mode is reduced from originally 41 to 29 hours and the period is estimated

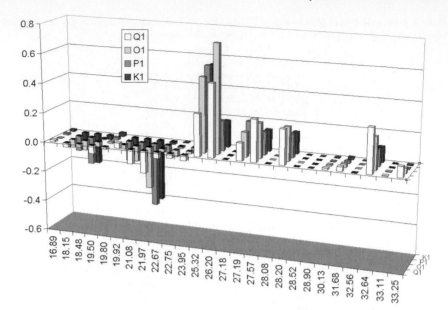

Fig. 4.23 The difference $\bar{a}_k - \bar{a}_k^{(a)}$ between the expansion coefficients $\bar{a}_k$ of the tidal synthesis of Section 4.1 and the newly obtained coefficients $\bar{a}_k^{(a)}$, when forced by the diurnal potentials of second degree (K_1, P_1, O_1, Q_1).

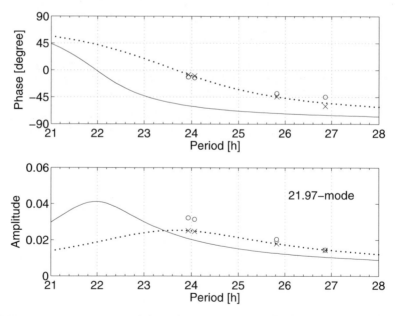

Fig. 4.24 The resonance curve $A_{\sigma_k}(\sigma)$ of the 21.97-mode according to equation (4.27). Information given as in Figure 4.18.

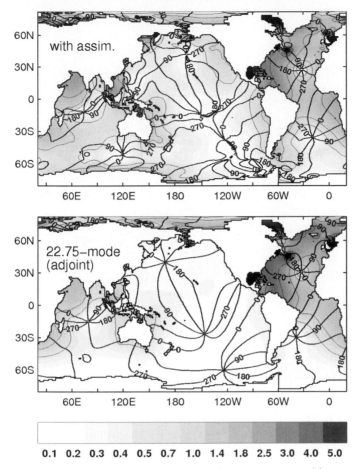

Fig. 4.25 The adjoint 22.75-mode (**bottom**) and the corresponding mode $\widehat{\mathbf{v}}_k^{(a)}$ (**top**) determined through minimizing (4.24) and (4.25). Gray shaded amplitudes of the sea surface elevation ζ in $[m]$. Solid lines showing the phases in degrees (in steps of $30°$).

with 28.6 hours.

The influence of the 27.19-mode is slightly overestimated, thus it keeps its dominant role for the O_1- and Q_1-tides. The remaining three important modes for the synthesis of the diurnal tides, the 27.57-, 28.20- and 32.64-mode are all overestimated, but the least squares fit is meaningless since their periods are too far from resonance.

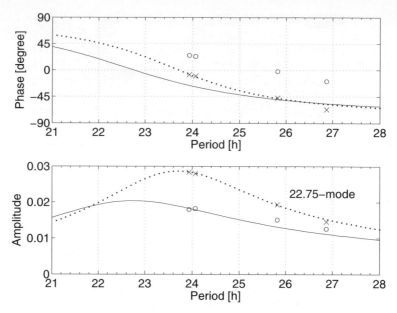

Fig. 4.26 The resonance curve $A(\sigma)$, equation (4.27), of the 22.75-mode. Information given as in Figure 4.18.

4.2.4 Summary

As described in Section 4.2.1, the syntheses utilizing the new expansion coefficients of (4.21) yield meaningful results. The obtained tidal patterns are in good agreement with the tidal solution of ZA2000. Previous analyzes of the tidal patterns showed that the tides are overestimated in the free tidal model (e.g. Zahel, 1995). This is reflected in the expansion coefficients, as well. The original expansion coefficients for the semidiurnal tides are all overestimated for K_2 and S_2. For these two constituents the changes are very large, especially for the two dominant modes (11.77- and 11.98-mode). For M_2 the smallest changes occur, mainly due to a significantly reduced influence of the 12.66-mode. The synthesis of N_2 also shows a reduced influence of the 12.66-mode and additionally significant underestimations of some modes with periods lower than 12.5 hours. The original expansion coefficients of the diurnal tides are mainly overestimated, as well. Only one of the important modes, the 21.97-mode is significantly underestimated.

The modifications of the expansion coefficients leads to a new order of relevance of the modes in the synthesis of the tides. For example the 11.89-mode loses its dominant role for S_2 and K_2, and the 12.66-mode for the M_2- and N_2-tides. This new synthesis yields that the 11.77-mode is the dominant mode for all four analyzed semidiurnal constituents. For the diurnal tides the order of the modes with respect to their importance changes, as well. For P_1 and K_1 the 21.97-mode replaces the 25.32- and 27.19-mode in their dominant role. In case of O_1 and Q_1 the 27.19-mode is solely dominant in contrast to the free model version; there the 26.20- and 27.19-mode are together the dominant modes.

The least squares fits of function (4.23) for selected modes in the semidiurnal range, yield period shifts between 0.06 and 0.42 hours and very large shifts of up to 26 hours for the damping rates. The adjustment of the adjoint eigenfunction of the 12.66-mode through the least squares fit reproduces the disappearance of the amphidrome south of Australia, as it is obtained by tidal models with assimilation of data. However, it is the adjoint solution and does not represent the contribution of the 12.66-mode to the tidal pattern. But the similarities of the 12.66-mode with its corresponding adjoint mode suggest that there is a connection between the disappearance of the amphidrome in semidiurnal tides with that one occurring for the adjoint 12.66-mode. The shifts of the periods of the modes in the diurnal range are all positive. They are very large amounting up to 4 hours.

4.2.5 Discussion

Using the new expansion coefficients influenced by tidal data, the synthesis of the tides yields considerably improved results as compared to those of Section 4.1.

Obtaining improved eigenfrequencies and adjoint eigenfunctions by using data of eight partial tides, as described in (4.2.2), is regarded as a first attempt. In particular in view of the small number of partial tides involved and the preliminary treatment of the nonlinear least squares problem, in fact, the frequency shift resulting from this attempt are questionable large, especially those of the damping rates. Further, the adjustment of the adjoint eigenfunction yields promising results only for one mode.

However, when further developing the presented method in different directions, this attempt to improve the description of the global free oscillations by means of additional data deserves being pursued.

Chapter 5
Conclusion

The objective of this study was to compute a large spectrum of free oscillations of a linear, barotropic, one degree global ocean model with explicit consideration of dissipative terms and the full LSA-effect. Concerning the LSA-effect in tides qualitative results are available for quite some time, while corresponding results for free oscillations were missing. The basic task consisted in introducing the LSA-effect in such a way that the resulting large scale eigenvalue problem would become accessible when applying a proper mathematical method.

It turned out that this task could be coped with by utilizing the Implicitly Restarted Arnoldi Method and implementing it in a specific way on a supercomputer. The ocean model is constructed to run parallel on a large number of CPUs of supercomputers with a shared memory distribution. The parallelization is required, to allow for the usage of the large amount of memory, needed for this kind of eigenvalue problem. With the developed ocean model the highly efficient computation of large spectra of free oceanic oscillations, with the full LSA-effect included, is enabled. Thus, the LSA-effect, which so far appears to be the greatest unknown left in the barotropic free oscillation behaviour of the open ocean, can now be better understood and quantitatively estimated. The following analyses have been performed:

A large spectrum of free oscillations has been computed applying the above mentioned global ocean model including the full LSA-effect and dissipative terms. To clarify the LSA-effect on barotropic ocean dynamics, an additional spectrum of free oscillations not influenced by LSA has been determined. This study of the effect of LSA on gravitational and vorticity modes of the World Ocean provides new insights into its influence on the eigenfrequency and on the fields of current velocity and sea surface elevation. Further, this study explains the difficulties with the parameterization of the LSA-effect in barotropic ocean models forced by tides, wind stress, and atmospheric pressure, since it turns out that the parameterization clearly depends for the gravity modes on the respective time scale. Furthermore, the approach by means of the Arnoldi Method yields a new group of free oscillations establishing a dense spectrum of topographical modes for periods longer than 13h. The extension of the spectrum towards periods longer than 4 days and up to 6.6 days reveals new global planetary modes and one planetary mode restricted to the Pacific Ocean.

For the free oscillations in the period range from 8 to 40 hours the corresponding adjoint solutions are also computed, in order to obtain a biorthonormal system of eigenvectors. A general procedure is derived to synthesize four diurnal and four semidiurnal tidal constituents from this system of eigenvectors. This approach allows for a detailed analysis of the expansion coefficients of each free oscillation. This analysis is performed firstly with respect to the resonance depth and the shape factor, and secondly with respect to the amplitude and phase of the expansion coefficients. Resonance depth and shape factor uniquely characterize the expansion coefficients, the former depending on the free oscillation period and on the forcing period, while the latter is defined through the correlation of the adjoint eigenvector with the forcing vector. The appearance of the delay of ocean tides by consideration of the LSA is physically explained. The delay is due to the influence of LSA on the phase of the resonance depths of near-resonant free oscillations. It is shown that this phase delay is much larger for the semidiurnal than for the diurnal tides. Further, a spectral composition in terms of free oscillations is performed of some well known tidal features, e.g. in the diurnal tides, the Antarctic Kelvin Wave and the topographical trapping of a vorticity mode south of New Zealand and in the semidiurnal tides the resonance in the Bay of Fundy - Gulf of Maine system.

Moreover, the synthesis procedure has been extended by a method including the solutions of a tidal model with assimilation of data. This approach yields new expansion coefficients for the free oscillations. A new order of relevance of the modes in the synthesis of the tides is found, and this analysis shows among other things that the contributions of the modes are predominantly overestimated in free tidal models. The data corrected coefficients lead to considerable improvement of tidal oscillation patterns. Moreover nonlinear least squares fits have been performed to compute improved frequencies and adjoint eigenvectors of certain free oscillations. The results of the least squares fits are still not satisfying and there are some improvements to be made in order to obtain successful results.

The latter attempt to combine the obtained free oscillations of the World Ocean with the results of ocean tide experiments with assimilation of data, in order to improve the description of the oceanic oscillation behaviour, should be continued. More than eight partial tides should be considered using this method and the large scale nonlinear least squares problem should be solved with a "full" nonlinear solver.

Furthermore, improvements in the quality of the free oscillations could be achieved through a refinement of the one degree grid. However, since the matrix is full the memory requirement would rise with $\mathcal{O}(N^3)$, N denoting the number of grid points, e.g. this results in an increase of the required working memory from around 600GByte to 9.6TByte, for a refinement to a half degree resolution. The access to supercomputers with this huge amount of working memory is currently problematic. Thus, a local refinement of the model grid would be preferable in order to save memory. The refined regions could comprise prominent topographical features in the open ocean making possible a better representation of topographical trapped vorticity modes.

Figures

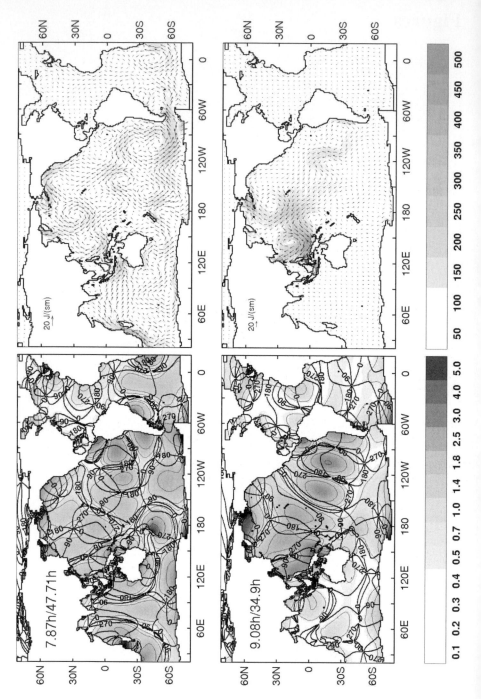

Fig. 5.1 Left: Normalized amplitudes of sea-surface elevations and lines of equal phases in degrees referred to 0° in (0.5°N, 89.5°W) (in steps of 45°). **Right:** Energy flux vectors shown for every fifth grid point, zonally and meriodionally. Squaring the magnitudes given yield these quantities in J/(sm). The energy flux vector at the top left has a magnitude of 20^2 J/(sm). The magnitude of the energy flux is additionally color contoured. **Top:** The 7.87-mode with a decay time of 47.71 h. **Bottom:** The 9.08-mode with a decay time of 34.90 h.

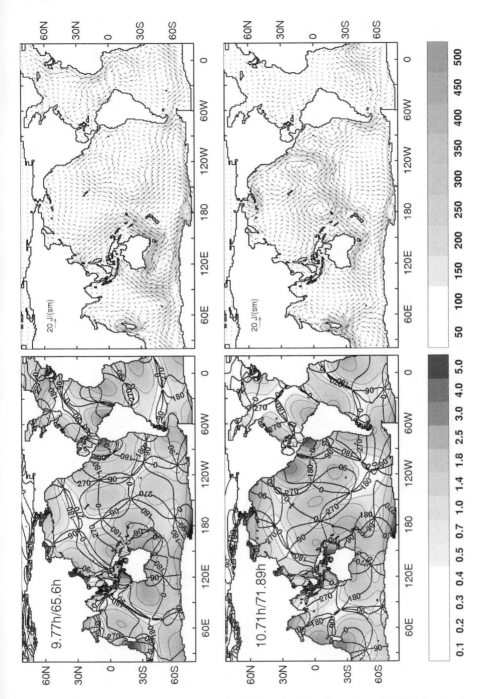

Fig. 5.2 Information given as in Figure 5.1. The 9.77-mode with a decay time of 65.60 h (**top**) and the 10.71-mode with a decay time of 71.89 h (**bottom**).

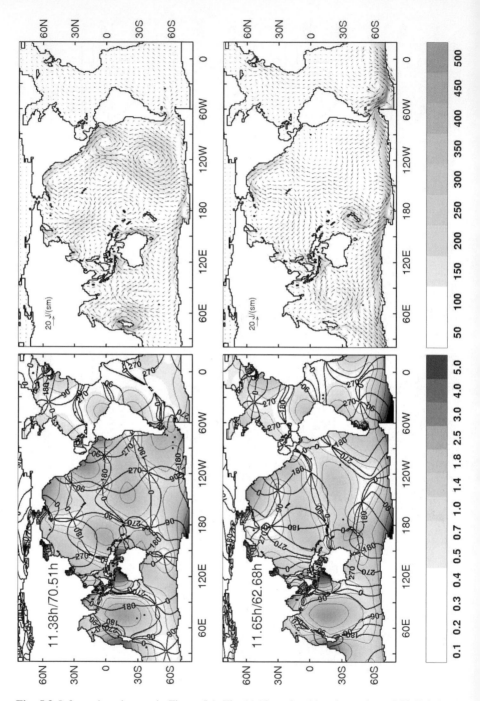

Fig. 5.3 Information given as in Figure 5.1. The 11.38-mode with a decay time of 70.51 h (**top**) and the 11.65-mode with a decay time of 62.68 h (**bottom**).

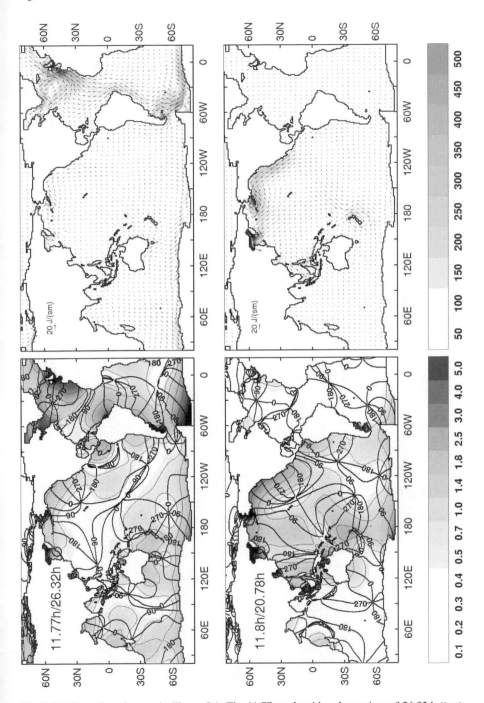

Fig. 5.4 Information given as in Figure 5.1. The 11.77-mode with a decay time of 26.32 h (**top**) and the 11.80-mode with a decay time of 20.78 h (**bottom**).

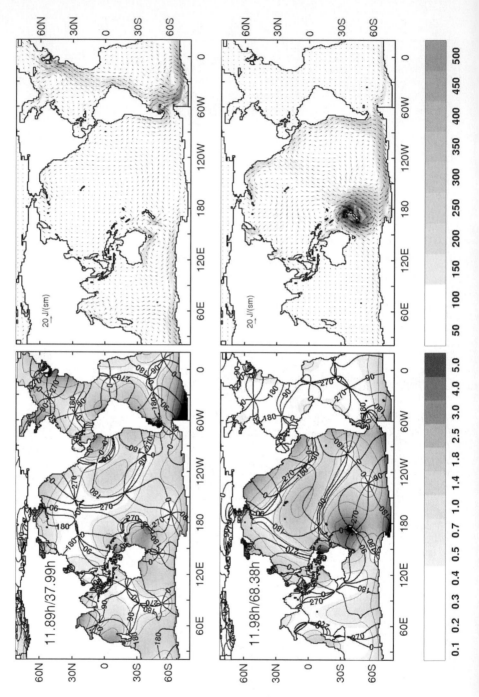

Fig. 5.5 Information given as in Figure 5.1. The 11.89-mode with a decay time of 37.99 h(**top**) and the 11.98-mode with a decay time of 68.38 h (**bottom**).

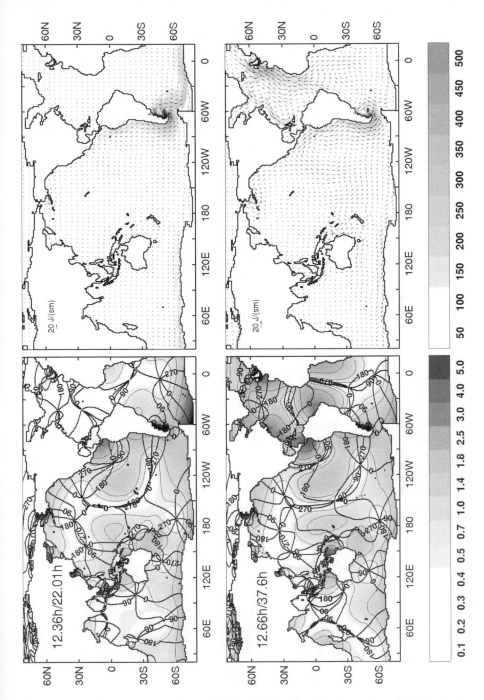

Fig. 5.6 Information given as in Figure 5.1. The 12.36-mode with a decay time of 22.01 h (**top**) and the 12.66-mode with a decay time of 37.6 h (**bottom**).

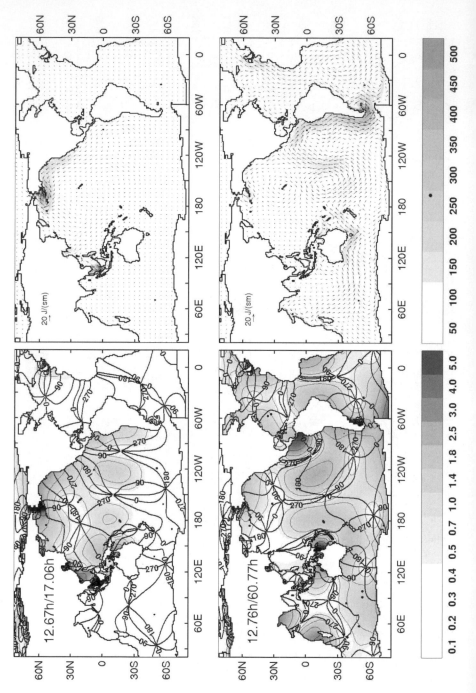

Fig. 5.7 Information given as in Figure 5.1. The 12.67-mode with a decay time of 17.06 h (**top**) and the 12.76-mode with a decay time of 60.77 h (**bottom**).

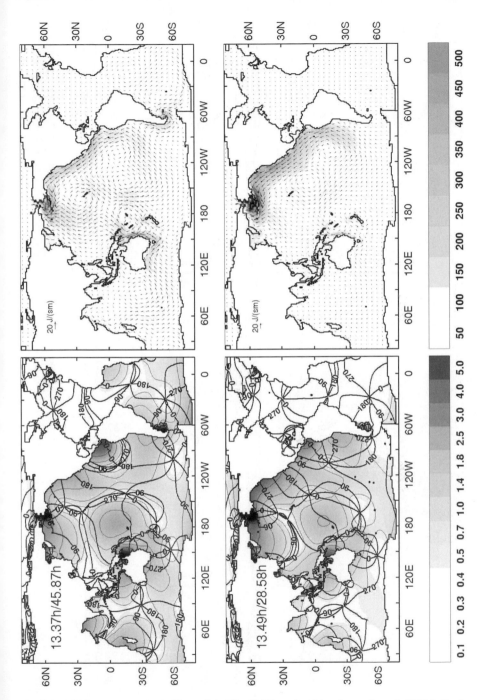

Fig. 5.8 Information given as in Figure 5.1. The 13.37-mode with a decay time of 45.87 h (**top**) and the 13.49-mode with a decay time of 28.58 h (**bottom**).

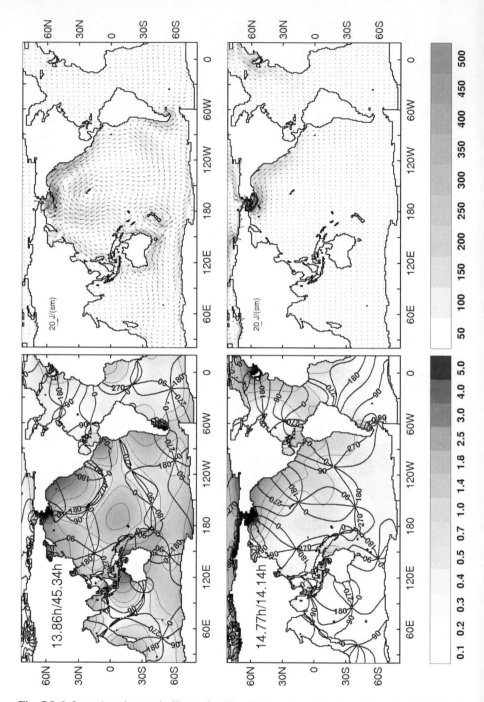

Fig. 5.9 Information given as in Figure 5.1. The 13.86-mode with a decay time of 45.34 h (**top**) and the 14.77-mode with a decay time of 14.14 h (**bottom**).

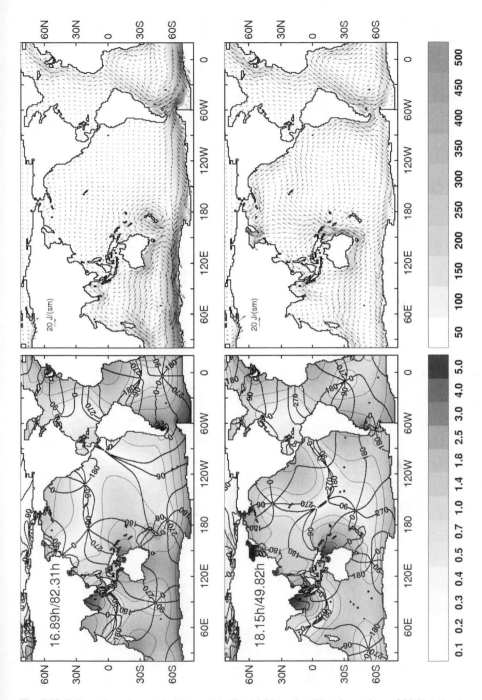

Fig. 5.10 Information given as in Figure 5.1. The 16.89-mode with a decay time of 82.31 h (**top**) and the 18.15-mode with a decay time of 49.82 h (**bottom**).

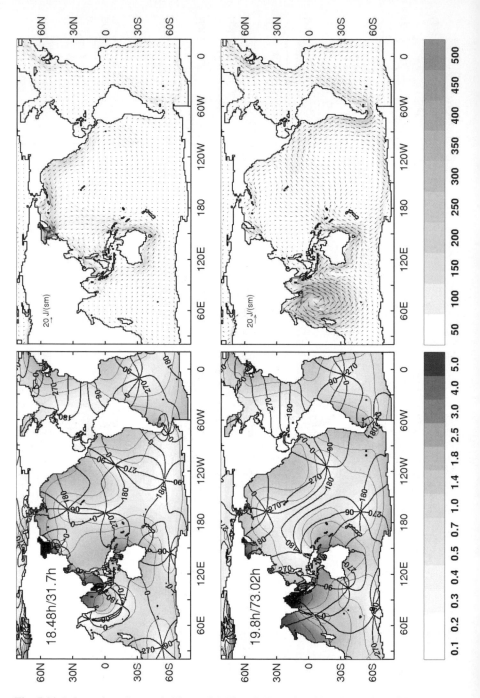

Fig. 5.11 Information given as in Figure 5.1. The 18.48-mode with a decay time of 31.7 h (**top**) and the 19.80-mode with a decay time of 73.02 (**bottom**).

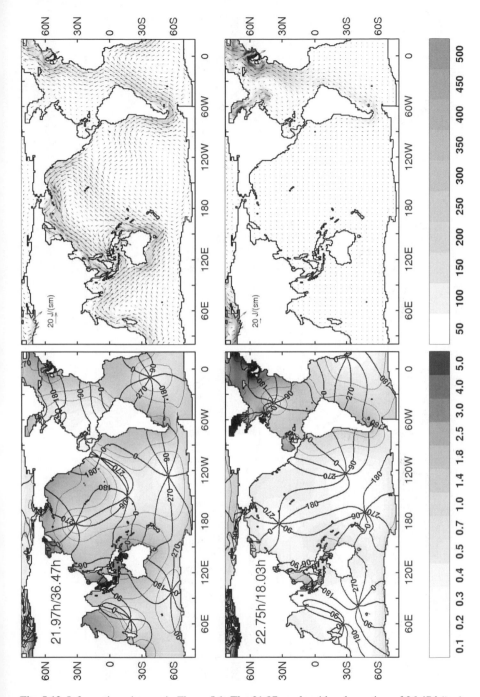

Fig. 5.12 Information given as in Figure 5.1. The 21.97-mode with a decay time of 36.47 h(**top**) and the 22.75-mode with a decay time of 18.03 h (**bottom**).

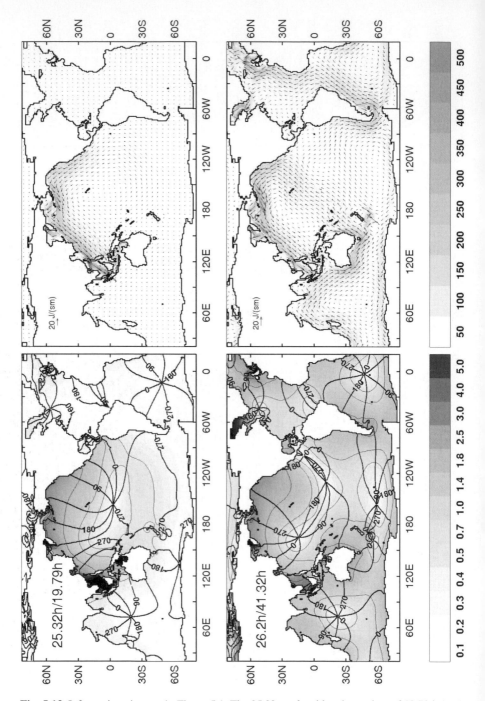

Fig. 5.13 Information given as in Figure 5.1. The 25.32-mode with a decay time of 19.79 h (**top**) and the 26.20-mode with a decay time of 41.32 h (**bottom**).

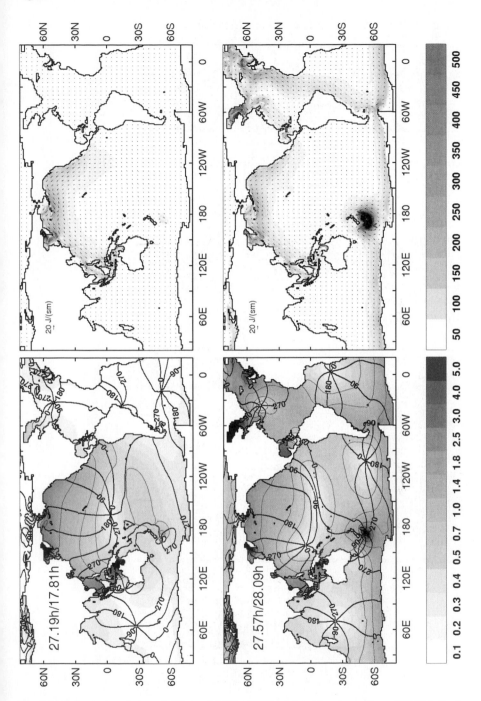

Fig. 5.14 Information given as in Figure 5.1. The 27.19-mode with a decay time of 17.81 h (**top**) and the 27.57-mode with a decay time of 28.09 h (**bottom**).

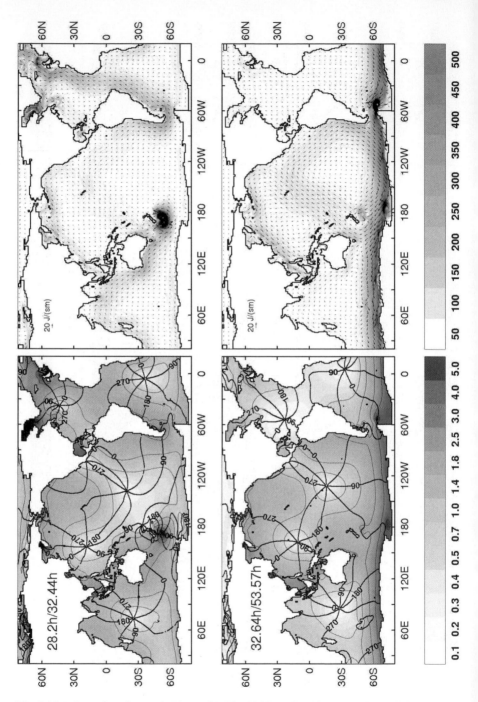

Fig. 5.15 Information given as in Figure 5.1. The 28.20-mode with a decay time of 32.44 h (**top**) and the 32.64-mode with a decay time of 53.57 h (**bottom**).

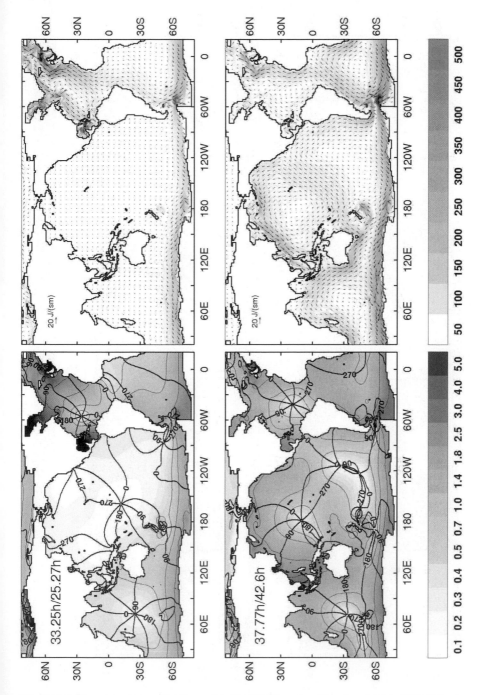

Fig. 5.16 Information given as in Figure 5.1. The 33.25-mode with a decay time of 25.27 (**top**) and the 37.77-mode with a decay time of 42.6 h (**bottom**).

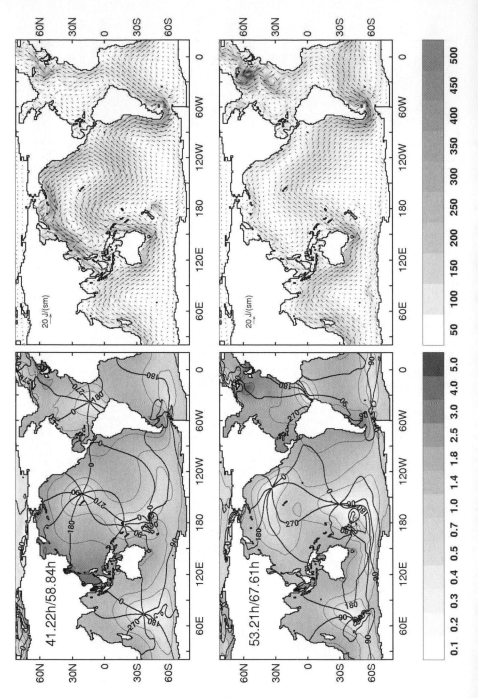

Fig. 5.17 Information given as in Figure 5.1. The 41.22-mode with a decay time of 58.48 h (**top**) and the 53.21-mode with a decay time of 67.61 h (**bottom**).

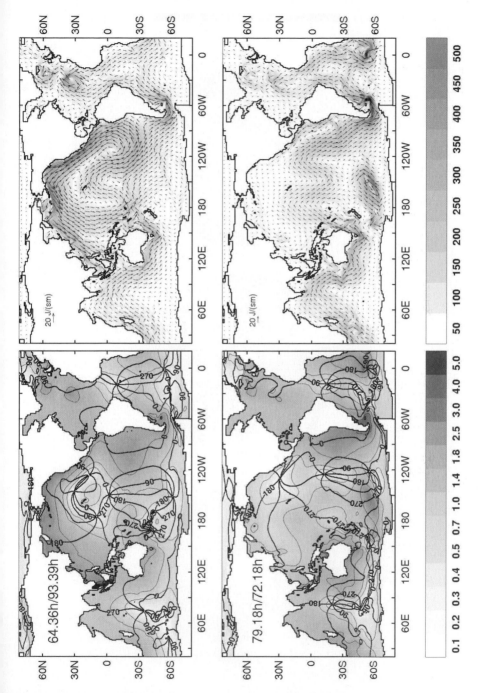

Fig. 5.18 Information given as in Figure 5.1. The 64.36-mode with a decay time of 93.39 h (**top**) and the 79.18-mode with a decay time of 72.18 h (**bottom**).

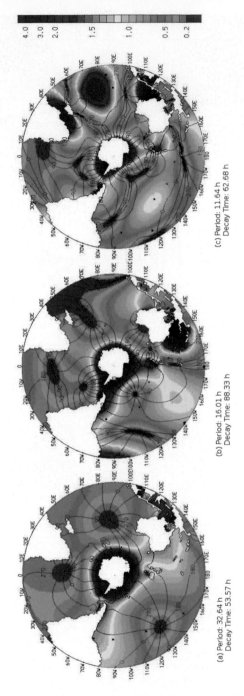

Fig. 5.19 Examples for Antarctic Kelvin Waves: (a) First, (b) second and (c.) third order Kelvin Waves travelling in counterclockwise direction around Antarctica. (Color contoured normalized amplitudes of sea-surface elevation and solid lines of equal phases).

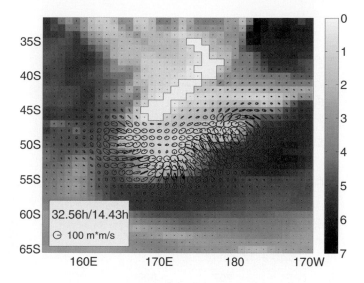

Fig. 5.20 Volume transport ellipses plotted for the topographical vorticity mode at the New Zealand Plateau with a period of 32.56 hours and a decay time of 14.43 hours. Red ellipses indicate cyclonic rotation, black ones anticyclonic rotation. The bathymetry is color-contoured and labeled in [km].

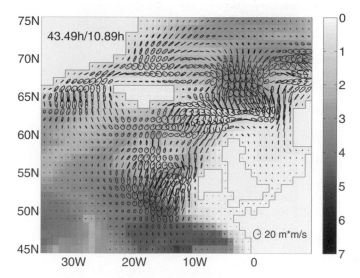

Fig. 5.21 Volume transport ellipses plotted for the topographical vorticity mode in the North Atlantic with a period of 43.49 hours and a decay time of 10.89 hours. Red ellipses indicate cyclonic rotation, black ones anticyclonic rotation. The bathymetry is color-contoured and labeled in [km].

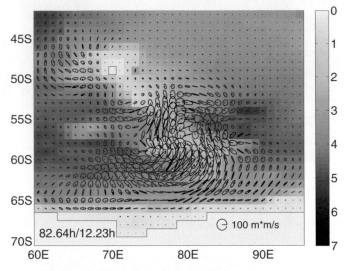

Fig. 5.22 Volume transport ellipses plotted for the topographical vorticity mode at the Kerguelen Plateau with a period of 82.64 hours and a decay time of 12.23 hours. Red ellipses indicate cyclonic rotation, black ones anticyclonic rotation. The bathymetry is color-contoured and labeled in [km].

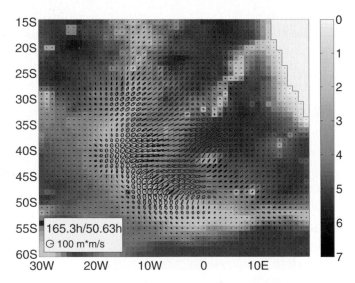

Fig. 5.23 Volume transport ellipses plotted for the topographical vorticity mode in the southern part of the Mid Atlantic Ridge with a period of 165.28 hours and a decay time of 50.63 hours. Red ellipses indicate cyclonic rotation, black ones anticyclonic rotation. The bathymetry is color-contoured and labeled in [km].

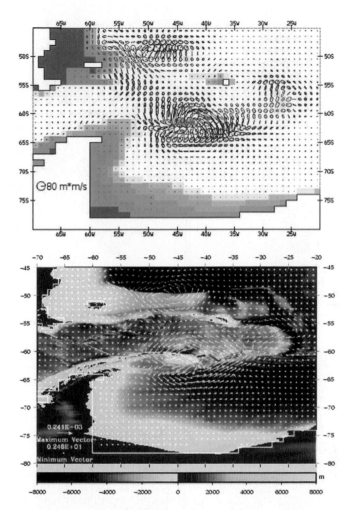

Fig. 5.24 Topographic vorticity mode around the Falkland Plateau and the South Orkney Islands with a period of 63.32 hours and an energy decay time of 12.23 hours (a) Volume transport ellipses. Blue contours indicating anticyclonic rotation, red ones cyclonic rotation, the circular volume transport given at the bottom left has a magnitude of 80 m^2/s. (b) Color contoured bathymetry. Energy flux vectors: squaring the magnitudes given, yields these quantities in J/(sm).

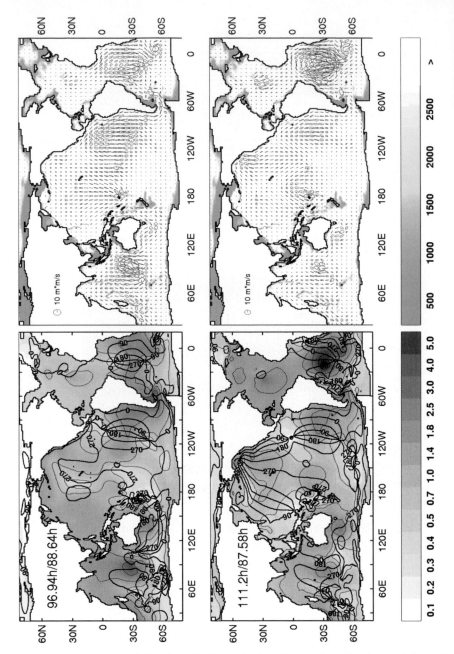

Fig. 5.25 Planetary Modes: **Left**: Normalized amplitudes of sea-surface elevations and lines of equal phases in degrees referred to 0° in (0.5°N, 89.5°W) (in steps of 45°). **Right**: Volume transport ellipses. Red ellipses indicate cyclonic rotation, blue ones anticyclonic rotation. The bathymetry is color-contoured and labeled in [km]. **Top**: The 96.94-mode with a decay time of 88.64 h. **Bottom**: The 111.22-mode with a decay time of 87.58 h.

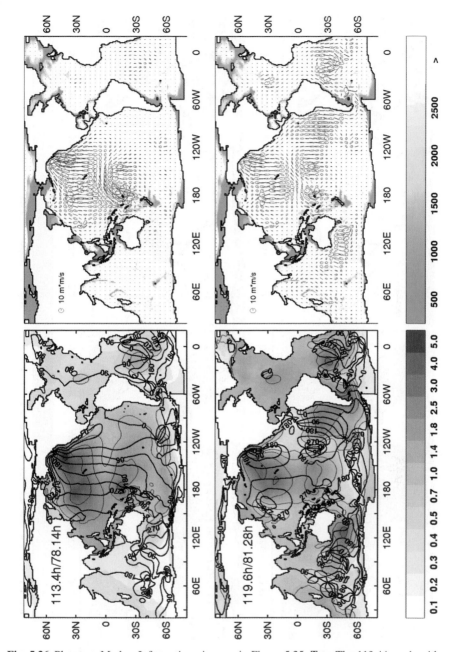

Fig. 5.26 Planetary Modes: Information given as in Figure 5.25. **Top**: The 113.44-mode with a decay time of 78.14 h. **Bottom**: The 119.65-mode with a decay time of 81.28 h.

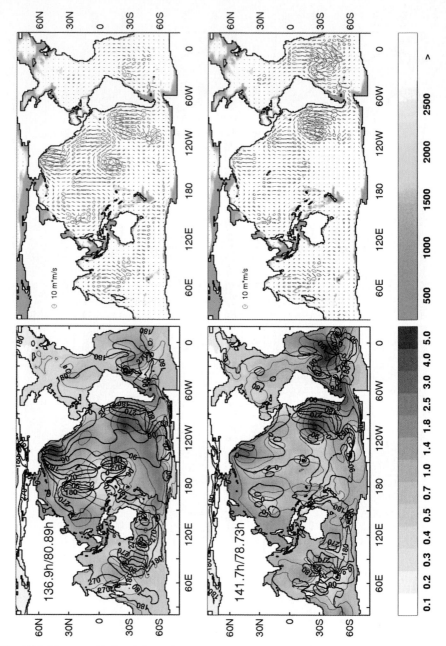

Fig. 5.27 Planetary Modes: Information given as in Figure 5.25. **Top**: The 136.87-mode with a decay time of 80.89 h. **Bottom**: The 141.66-mode with a decay time of 78.73 h.

Tables

Table 5.1 The gravitational normal modes in the period range 7.77-11.98 hours. The period $T_2 = \frac{2\pi}{\sigma_2}$, the decay time $T_1 = \frac{1}{2\sigma_1}$, the period T_2^p of the corresbonding mode computed with parameterized LSA (if it exists), and the ratio between potential (E_p) and total (E_t) energy contents is shown for each mode. Further, areas relative to the ocean area are given in per cent for Indian Ocean (Ind), Pacific Ocean (Pac), Atlantic Ocean (Atl), Southern Ocean (Ant) and the North Polar Sea (Np) directly below the abbrevations. The five columns below these abbrevations show the total energy content of the corresponding ocean region relative to the total energy. The last column shows the β-value times 100.

No.	$T_2[h]$	$T_2^p[h]$	$T_1[h]$	$\frac{E_p}{E_t}$ [%]	Ind 18.7	Pac 46.6	Atl 23.2	Ant 8.5	Np 3.0	$\beta \cdot 100$
1	7.77		38.91	47.4	27.1	40.2	18.2	14.2	0.3	5.41
2	7.81		28.56	49.5	53.6	17.6	22.3	5.9	0.6	4.83
3	7.87		47.71	46.9	24.1	50.9	12.3	12.7	0.1	5.91
4	7.91	8.03	21.74	49.9	6.5	7.6	79.9	5.8	0.2	4.34
5	8.06	8.12	55.60	46.7	10.0	51.7	15.5	22.7	0.1	5.83
6	8.08		21.24	48.2	4.2	24.2	61.2	9.6	0.8	4.14
7	8.15	8.22	36.24	47.3	4.5	71.9	6.8	16.8	0.1	5.64
8	8.16		19.62	49.0	2.0	94.1	1.7	2.1	0.0	4.50
9	8.26		15.07	49.8	4.1	5.5	81.9	3.4	5.2	2.82
10	8.29	8.41	30.18	48.5	10.6	31.5	37.6	19.0	1.2	4.78
11	8.37		28.46	49.3	8.9	28.0	45.4	15.6	2.1	4.73
12	8.47		24.21	50.3	6.5	12.7	64.4	12.6	3.8	4.29
13	8.66	8.72	57.52	47.3	2.0	86.4	2.8	8.8	0.0	6.63
14	8.74	8.88	29.19	49.2	22.1	8.9	53.7	12.3	3.1	4.93
15	8.88	8.98	49.75	48.7	9.6	57.5	15.3	17.0	0.6	5.83
16	8.98		31.13	49.1	16.6	7.5	59.4	14.0	2.5	5.08
17	9.08	9.12	34.90	47.0	3.5	87.6	4.9	3.7	0.2	6.41
18	9.22	9.25	87.21	43.5	8.1	60.7	7.6	23.4	0.2	6.89
19	9.42	9.50	42.69	46.1	58.1	14.6	17.9	8.9	0.4	6.23
20	9.42		16.86	49.3	1.0	98.2	0.1	0.6	0.0	3.82
21	9.43		67.21	44.8	17.0	54.6	11.2	17.0	0.2	6.89
22	9.66		15.83	50.0	1.8	97.3	0.1	0.7	0.1	3.55
23	9.77	9.82	65.60	46.5	27.6	36.4	23.5	11.9	0.5	6.79
24	9.82		18.87	48.9	2.1	96.9	0.2	0.8	0.1	4.29
25	9.82		24.46	49.7	1.3	10.5	79.6	1.5	7.0	4.96
26	9.94	10.02	33.17	44.3	3.4	6.9	78.5	11.0	0.2	5.46
27	10.27	10.32	96.52	45.2	40.4	20.0	30.8	8.8	0.0	7.31
28	10.71	10.75	71.89	46.9	20.2	55.8	16.4	7.6	0.1	7.37
29	10.88	10.87	35.10	47.2	8.6	87.3	1.9	2.2	0.0	6.78
30	10.98	11.03	66.09	46.7	39.9	35.2	12.0	12.8	0.1	7.20
31	11.38	11.37	70.51	44.3	30.1	56.7	4.7	8.5	0.1	7.83
32	11.65	11.65	62.68	45.7	33.0	33.5	10.2	23.1	0.2	6.94
33	11.77	11.88	26.32	47.6	6.3	16.1	60.2	14.4	3.0	5.17
34	11.80		20.78	50.4	2.0	94.2	0.9	2.8	0.1	4.28
35	11.89	11.96	37.99	45.9	14.6	24.7	40.7	18.9	1.1	5.86
36	11.98		68.38	46.6	3.5	79.9	1.8	14.7	0.0	7.49

Table 5.2 The gravitational normal modes in the period range 12.36-79.18 hours. Remaining information given as in Table A.1.

No.	$T_2[h]$	$T_2^p[h]$	$T_1[h]$	$\frac{E_p}{E_t}$ [%]	Ind 18.7	Pac 46.6	Atl 23.2	Ant 8.5	Np 3.0	$\beta \cdot 100$
37	12.36		22.01	47.0	2.9	26.5	59.8	10.8	0.1	4.30
38	12.55		13.88	48.9	0.3	93.2	1.0	0.1	5.5	3.50
39	12.66	12.65	37.60	47.1	4.3	32.1	56.4	5.6	1.5	6.99
40	12.67		17.06	50.0	0.2	97.3	0.8	0.1	1.6	3.57
41	12.76	12.75	60.77	45.3	14.9	50.6	26.6	7.5	0.4	7.77
42	13.37	13.39	45.87	44.6	14.5	73.0	9.1	3.0	0.3	7.71
43	13.49		28.58	46.0	4.6	92.5	0.9	0.9	1.2	6.64
44	13.86	13.92	45.34	45.5	14.0	73.1	7.9	4.6	0.5	7.62
45	14.38		31.88	47.3	5.3	17.9	69.1	3.8	3.8	6.00
46	14.60	14.55	34.81	46.3	11.1	23.6	57.9	4.4	2.9	6.15
47	14.77	14.76	14.14	43.8	0.1	66.9	10.4	0.1	22.4	3.49
48	15.36	15.38	57.69	45.7	27.4	35.6	27.2	8.4	1.4	8.07
49	16.02	15.98	88.33	47.9	9.4	55.1	5.3	29.8	0.4	8.48
50	16.39	16.31	33.63	48.0	6.0	89.9	2.1	1.1	0.9	7.12
51	16.89	16.81	82.31	44.1	29.3	19.5	21.8	27.6	1.8	8.97
52	18.15	18.04	49.82	46.7	22.5	49.0	17.0	8.3	3.2	8.25
53	18.48		31.70	47.7	11.9	78.1	5.8	2.1	2.2	6.43
54	19.50	19.61	20.30	42.4	0.2	5.3	46.8	1.1	46.6	5.32
55	19.80		73.02	45.4	64.5	24.6	6.4	3.1	1.4	9.03
56	19.92	20.26	23.07	50.7	1.2	97.9	0.3	0.1	0.5	4.90
57	21.08	21.21	30.76	49.9	14.4	73.5	8.2	0.9	3.0	7.47
58	21.97	21.71	36.47	49.3	17.6	58.2	17.4	2.2	4.6	8.57
59	22.75	23.00	18.03	39.3	0.3	3.7	67.7	0.9	27.4	5.13
60	25.32		19.79	50.6	1.0	98.2	0.5	0.1	0.2	6.61
61	26.20	25.76	41.32	44.7	10.9	57.7	23.7	4.4	3.3	10.17
62	27.19		17.81	46.7	1.2	98.3	0.2	0.2	0.2	5.43
63	27.57	27.66	28.09	26.0	0.9	66.9	22.0	6.5	3.6	7.76
64	28.20	28.15	32.44	27.3	6.7	58.2	26.2	4.9	4.0	9.24
65	32.64	31.73	53.57	43.4	8.0	39.0	4.2	48.5	0.3	11.91
66	33.25	33.26	25.27	40.1	2.1	2.1	72.4	14.7	8.7	8.34
67	37.77	37.05	42.60	36.0	12.2	41.0	21.0	24.4	1.4	11.50
68	41.22	40.01	58.84	42.2	9.9	66.3	18.6	4.1	1.1	13.39
69	53.21	51.72	67.61	34.1	13.0	33.2	45.7	6.2	1.9	14.21
70	64.36	63.57	93.39	20.6	6.1	74.3	17.1	2.2	0.3	13.54
71	79.18	77.82	72.18	18.6	16.2	41.0	33.0	9.7	0.1	14.39

Appendix

Phase Delay

The resonance depth of the kth-mode is defined as $R_k = \frac{1}{\sigma_{k,1} + i(\sigma_{k,2} - \sigma_F)}$. We can rewrite the resonance depth in the form:

$$R_k = \frac{1}{\sqrt{\sigma_{k,1}^2 + (\sigma_{k,2} - \sigma_F)^2}} \cdot e^{i \cdot \phi_k}. \tag{5.1}$$

The phase ϕ_k is given by $\phi_k = arctan\left(-\frac{\sigma_{k,2} - \sigma_F}{\sigma_{k,1}}\right)$. With the assumption that for the phase delay induced by the LSA, the change in the oscillatory frequency part $\delta\sigma_k = \sigma_{k,2}^{(LSA)} - \sigma_{k,2}^{(noLSA)}$ plays the major role, we set $\delta\sigma_k = \sigma_{k,1}^{(LSA)} - \sigma_{k,1}^{(noLSA)} = 0$. Thus, we obtain with the identity $arctan(x) - arctan(y) = arctan\left(\frac{x-y}{1+xy}\right)$:

$$\delta\phi_k = \phi_k^{(LSA)} - \phi_k^{(noLSA)} =$$
$$arctan\left(\frac{\sigma_{k,2}^{(noLSA)} - \sigma_{k,2}^{(LSA)}}{\sigma_{k,1}\left(1 + (\sigma_{k,2}^{(noLSA)} - \sigma_F)(\sigma_{k,2}^{(LSA)} - \sigma_F)/\sigma_{k,1}^2\right)}\right) \tag{5.2}$$

List of Symbols

a		angular distance
a_k		expansion coefficient of the k-th free oscillation
$a_k^{(a)}$		assimilated expansion coefficient of the k-th free oscillation
A		complex Matrix of dimension $n \times n$
A_h	$[m^2 s^{-1}]$	kinematic eddy viscosity for the horizontal direction
α_n		normalized density ratio
β		global measure for the LSA effect
β_L		local measure for the LSA effect
C_k		shape factor of the k-th free oscillation
D	$[m]$	undisturbed ocean depth
δ	$[m]$	elevation of sea bottom
$\mathbf{e}_l$		l-th unit vector of length n
$\overline{E}_k$	$[Jm^{-2}]$	mean kinetic energy
$\overline{E}_p$	$[Jm^{-2}]$	mean potential energy
$\overline{E}_t$	$[Jm^{-2}]$	mean total energy
$\mathbf{f}$	$[s^{-1}]$	vector of Coriolis acceleration
$\mathbf{F}$	$[ms^{-2}]$	vector of second-order eddy viscosity term
$\tilde{\mathbf{F}}$		external tidal forcing
ϕ	$[deg]$ or $[rad]$	geographic latitude
Φ	$[m^2 s^{-2}]$	potential due to the self-attraction effect
Φ^*	$[m^2 s^{-2}]$	potential due to the loading and self-attraction effect
$\delta\phi_k$	$[deg]$ or $[rad]$	LSA induced phase delay
g	$[ms^{-2}]$	surface gravity of a spherical earth

G		Green-function of loading and self-attraction
γ	$[m^3 s^{-2} kg^{-1}]$	gravitational constant
h_n, k_n		nonloading Love numbers
h_n', k_n'		loading Love numbers
H_k		upper Hessenberg matrix
$\bar{J}_u$	$[J(ms)^{-1}]$	zonal time-mean energy flux
$\bar{J}_v$	$[J(ms)^{-1}]$	meridional time-mean energy flux
δK_R		spherical layer of radius R
$\mathcal{K}_k$		k-th Krylov subspace
$\mathbf{L}_{sek}$	$[ms^{-2}]$	vector of secondary force of the loading and self-attraction
$\mathcal{L}$		Operator derived through the shallow water equations
$\mathcal{L}_0$		Operator $\mathcal{L}$ without LSA-term
λ	$[deg]$ or $[rad]$	geographic east longitude
ω	$[s^{-1}]$	rotational angular velocity of the earth
$\bar{P}_{n,s}$		normalized Legendre polynomials of degreen n and order s
r		residual for the eigen-pair
r'	$[ms^{-1}]$	coefficient of linear bottom friction
R	$[m]$	mean radius of the earth
R_k	$[s]$	resonance depth of the k-th free oscillation
ρ	$[kgm^{-3}]$	density of sea water
ρ_O	$[kgm^{-3}]$	mean density of sea water
ρ_e	$[kgm^{-3}]$	mean density of solid earth
t	$[s]$	time
u	$[ms^{-1}]$	zonal current velocity
$\mathbf{u}_l$		l-th basis vectors of $\mathcal{K}_k$ with length n
v	$[ms^{-1}]$	meridional current velocity
$\mathbf{v}$	$[ms^{-1}]$	horizontal current velocity vector
$\mathbf{v}_k$		k-th eigenfunction
$\widehat{\mathbf{v}}_k$		k-th adjoint eigenfunction
$\widehat{\mathbf{v}}_k^{(a)}$		assimilated k-th adjoint eigenfunction
$\mathbf{x}$		eigenvector of A
θ	$[s^{-1}]$	Ritz value
σ	$[s^{-1}]$	complex eigenfrequency of free oscillation

σ_0	$[s^{-1}]$	first-guess eigenfrequency
σ_1	$[s^{-1}]$	damping rate of free oscillation
σ_2	$[s^{-1}]$	oscillationary part of the complex eigenfrequency σ
$\sigma_k^{(a)}$	$[s^{-1}]$	assimilated frequency of k-th free oscillation
T_1	$[s]$	damping rate of free oscillation
$T_1^{(a)}$	$[s]$	assimilated damping rate of free oscillation
T_2	$[s]$	period of free oscillation
$T_2^{(a)}$	$[s]$	assimilated period of free oscillation
ζ	$[m]$	sea surface elevation
$\overline{\zeta}$	$[m]$	equilibrium tide of the secondary potential of LSA
ζ_n	$[m]$	n-th degree spherical harmonic constituent of ζ
ζ_0	$[m]$	geocentric sea surface elevation
ζ_*	$[m]$	complex conjugate of ζ

References

1. Accad, Y., Pekeris, C.L.: Solution of the tidal equations for the M_2 and S_2 tides in the world oceans from a knowledge of the tidal potential alone. Phil. Trans. Roy. Soc., London **A290**, 235–266, (1978).
2. Arnoldi, W.E.: The principle of minimized iterations in the solution of the matrix eigenvalue problem. Quaterly of Applied Mathematics **9**, 17–29, (1951).
3. Bartels, J.: Handbuch der Physik, Volume Band XLVIII of Geophysik II, Chapter Gezeitenkräfte, (1957).
4. Blackford, L.S., Choi, J., Cleary, A., D'Azevedo, E., Demmel, J., Dhillon, I., Dongarra, J., Hammarling, S., Henry, G., Petitet, A., Stanley, K., Walker, D., Whaley, R.C.: Scalapack users' guide. SIAM, (1997).
5. Dziewonski, A.M., Anderson, D.L.: Preliminary reference earth model. Phys. Earth Planet. Interior **25**, 297–356, (1981).
6. Egbert, E.D., Ray, R.D.: Estimates of M_2 tidal energy dissipation from Topex/Poseidon altimeter data. J. Geophys. Res. **106**, (2001).
7. Egbert, E.D., Ray, R.D., Bills, B.G.: Numerical modeling of the global semidiurnal tide in the present and in the last glacial maximum. J. Geophys. Res. **109**, (2004).
8. Estes, R.H.: A computer software system for the generation of global ocean tide including self-attraction and ocean loading effects. Technical Report Report No. X-920-77-8, NASA/Goddard Space Flight Center, (1977).
9. Farrell, W.E.: Deformation of the earth by surface loads. Rev. Geophys. Space Phys. **10**, 761–797, (1972).
10. Francis, O., Mazzega, P.: Global charts of of ocean tide loading effects. J. Geophys. Res. **95**, 411–424, (1990).
11. Garrett, C.: Tidal resonance in the bay of fundy and gulf of maine. Nature **238**, 441–443, (1972).
12. Gaviño, J.H.R.: Über die Bestimmung von reibungslosen barotropen Eigenschwingungen des Weltozeans mittels der Lanczos-Methode. PhD thesis, Univ. Hamburg, (1981).
13. Golub G., V. Loan, C.: Matrix Computations - 2nd Edition. Johns Hopkins University Press, (1989).
14. Gotlib, V.Y., Kagan, B.A.: On the resonance excitation of semidiurnal tides in the world ocean. Fizika atmosfery i okeana **17**, 502–512, (1981).
15. Gotlib, V.Y., Kagan, B.A., Kivman, G.A.: On the lowest gravitational mode of eigen oscillations in the World Ocean: Part 2. Sov. J. Phys. Oceanogr **1**, 11–16, (1987).
16. Gordeev, R.G., Kagan, B.A., Polyakov, E.V.: The effects of loading and self-attraction on global ocean tides: the model and results of a numerical experiment. J. Phys. Oceanogr, **7**, 161–170, (1977).
17. Ipsen, I.C.F.: A history of inverse iteration. In H. S. B. Huppert, editor, Helmut Wielandt, Mathematische Werke, pages 464–472. Walter de Gruyter, (1996).

18. Kowalik, Z.: Modeling of topographically amplified diurnal tides in the nordic seas. J. Phys. Oceanogr. **24**, 1717–1731, (1994).
19. Lanczos, C.: An iteration method for the solution of the eigenvalue problem of linear differential and integral operators. Journal of Research of the National Bureau of Standards **45**, 255–282, (1950). Research Paper 2133.
20. Lehoucq, R.B., Sorensen, D.C., Yang, C.: ARPACK user's guide: Solution of large scale eigenvalue problems by implicitly restarted arnoldi methods. Technical report, Rice University, Department of Computational and Applied Mathematics, (1996).
21. Longuet-Higgins, M.S.: The eigenfunctions of laplace's tidal equations over a sphere. Phil. Trans. Roy. Soc. London **A262**, 511–581, (1968).
22. Longuet-Higgins, M.S., Pond, G.S.: The free oscillations of fluid on a hemisphere bounded by meridians of longitude. Phil. Trans. Roy. Soc. London **A266**, 193–223, (1970).
23. Marchuk, G.I., Kagan, B.A.: Dynamics of Ocean Tides, Kluwer Academic Publishers, (1989).
24. Lamb, H.: Hydrodynamics. Cambridge at the university press, (1932).
25. Levenberg, K.: A method for the solution of certain non-linear problems in least squares. Quart. Appl. Math., (1944).
26. Luther, D.S.: Evidence of a 4-6 day barotropic, planetary oscillation of the Pacific Ocean. J. Phys. Oceanogr. **12**, 644–657, (1982).
27. Miller, A.J.: On the barotropic planetary oscillations of the pacific. J. Mar. Res. **47**, 569–594, (1989).
28. Miller, A.J., Lermusiaux, P.F.J., Poulain, P.:A topographic-rossby mode resonance over the iceland-faeroe ridge.J. Phys. Oceanogr. **26**, 2735–2747, (1996).
29. Müller, M.: Barotrope Eigenschwingungen im Globalen Ozean unter Berücksichtigung des vollen Selbstanziehungs- und Auflasteffektes. Master's thesis, University Hamburg, (2003).
30. Müller, M.: The free oscillations of the world ocean in the period range 8 to 165 hours including the full loading effect. Geophys. Res. Lett. **34** (L05606), (2007).
31. Müller, M.: Synthesis of forced oscillations, Part 1: Tidal dynamics and the influence of LSA. Ocean Modelling **20**, 207–222, (2008).
32. Paige, C.C., Saunders, M.A.: LSQR: An algorithm for sparse linear equations and sparse least squares. ACM Transactions on Mathematical Software **8**, 43–71, (1982).
33. Parke, M.E.: O_1,P_1,N_2 models of the global ocean tide on an elastic earth plus surface potential and spherical harmonic decompositions for M_2,S_2 and K_1. Mar. Geod. **6**, 35–81, (1982).
34. Platzman, G.W.: Normal modes of the World Ocean, Part 3: A procedure for tidal synthesis. J.Phys.Oceanogr. **14**, 1521–1531, (1984).
35. Platzman, G.W: Normal modes of the Wworld Ocean, Part 4: Synthesis of diurnal and semidiurnal tides. J.Phys.Oceanogr. **14**, 1532–1550, (1984).
36. Platzman, G.W.: Tidal evidence for ocean normal modes. In B. B. Parker, editor, Tidal Hydrodynamics, pages 13–26, John Wiley and Sons, (1991).
37. Platzman, G.W., Curtis, G.A., Hansen, K.S., Slater, R.D.: Normal modes of the World Ocean, Part 2: Description of modes in the range 8 to 80 hours, J.Phys.Oceanogr. **11**, 579–603, (1981).
38. Ponte, R.M.: Nonequilibrium response of the global ocean to the 5-day Rossby-Haurwitz wave in atmospheric surface pressure. J. Phys. Oceanogr. **27**, 2158–2168, (1997).
39. Ponte, R.M., Hirose, N.: Propagating bottom pressure signals around Antarctica at 1-2-periods and implication for ocean modes. J.Phys.Oceanogr. **34**, 284–292, (2004).
40. Rao, D.B.: Free gravitational oscillations in rotating rectangular basins. J. Fluid Mech. **25**, 523–555, (1966).
41. Ray, R.D.: Ocean self-attraction and loading in numerical tidal models. Mar. Geod. **21**, 181–192, (1998).
42. Ray, R.D.: Resonant third-degree diurnal tides in the seas off western europe. J. Phys. Oceanogr. **31**, 3581–3586, (2001).
43. Saad, Y.: Numerical Methods for Large Eigenvalue Problems: Theory and Algorithms. John Wiley, New York, (1992).

44. Shum, C.K., Woodworth, P.L., Andersen, O.B., Egbert, G.D., Francis, O., King, C., Klosko, S.M., LeProvost, C., Li, X., Molines, J.M., Parke, M.E., Ray, R.D., Schlax, M.G., Stammer, D., Tierney, C.C. Vincent, P., Wunsch, C.I.: Accuracy assessment of recent ocean tide models. J. Geophys. Res. **102** (C11), (1997).

45. Smirnow, W.I.: Lehrgang der höheren Mathematik III,2. VEB Deutscher Verlag der Wissenschaften, (1964).

46. Snir, M., Otto, S., Huss-Lederman, S., Walker, D., Dongarra, J.: MPI - The Complete Reference Vol. 1: The MPI core. MIT Press, (1998).

47. Sorensen, D.C.: Implicit application of polynomial filters in a k-step Arnoldi method. SIAM Journal on Matrix Analysis and Applications **13**, 357–385, (1992).

48. Stepanov, V.N., Hughes, C.W.: Parameterization of ocean self-attraction and loading in numerical models of the ocean circulation. J. Geophys. Res. **109**, C03037, (2004).

49. Sutherland, G., Garrett, C., Foreman, M.: Tidal resonance in Juan de Fuca Strait and the Strait of Georgia. J. Phys. Oceanogr. **35**, 1279–1286, (2005).

50. Thomas, M.: Ergebnisse eines Simultanmodells für Zirkulation und ephemeridische Gezeiten im Weltozean. PhD thesis, Univ. Hamburg, (2001).

51. Walters, R.A. Goring, D.G.: Ocean tides around New Zealand. New Zealand Journal of Marine and Freshwater Research **35**, 567–579, (2001).

52. Webb, D.J.: Green's function and tidal prediction. Rev. Geophys. Space Phys. **12**, 103–116, (1974).

53. Weijer, W., Gille, S.T.: Adjustment of the southern ocean to wind forcing on synoptic time scales. J.Phys.Oceanogr. **35**, 2076–2089, (2005).

54. Weis, P.: Ocean Tides and the Earths Rotation - Results of a High-Resolving Ocean Model forced by the Lunisolar Tidal Potential. PhD thesis, Univ. Hamburg, (2006).

55. Wielandt, H.: Das Iterationsverfahren bei nicht selbstadjungierten linearen Eigenwertaufgaben. Math. Z. **50**, 93–143, (1944).

56. Zahel, W.: The influence of solid earth deformations on semidiurnal and diurnal ocean tides. In P. Brosche and J. Sündermann, editors, Tidal friction and the Earth's rotation, pages 98–124. Springer-Verlag, (1978).

57. Zahel, W.: Mathematical modelling of global interaction between ocean tides and earth tides. Phys. Earth Planet. Interior **21**, 202–217, (1980).

58. Zahel,W.: Astronomical tides. In J. Sündermann, editor, Numerical Data and Functional Relationships in Science and Technology, Volume 3c of Landolt-Börnstein, Chapter 6.4, pages 83–134. Springer-Verlag, (1986).

59. Zahel, W.: The influence of ocean and solid earth parameters on oceanic eigenoscillations, tides and tidal dissipation. Geophys. Monograph **59**, (1990a).

60. Zahel, W.: The consideration of solid earth effects in ocean tide modeling. In P. Brosche and J. Sündermann, editors, Earth's Rotation from Eons to Days, pages 69–80. Springer-Verlag, (1990b).

61. Zahel, W.: Modeling ocean tides with and without assimilating data. J.Geophys.Res. **12**, 20379–20391, (1991).

62. Zahel, W.: Assimilating ocean tide determined data into global tidal models. J.Mar.Syst., 1994.

63. Zahel, W., Gaviño, J.H. Seiler, U.: Balances de energia y momento angular de un modelo global de mareas con asimilacion de datos. GEOS, **4**, 400–413, (2000).

64. Zahel, W., Müller, M.: The computation of the free barotropic oscillations of a global ocean model including friction and loading effects. Ocean Dynamics **55**, 137–161, (2005).

Index

About the International Max Planck Research School for Maritime Affairs at the University of Hamburg

The International Max Planck Research School for Maritime Affairs at the University of Hamburg was established by the Max Planck Society for the Advancement of Science, in co-operation with the Max Planck Institute for Foreign Private Law and Private International Law (Hamburg), the Max Planck Institute for Comparative Foreign Public Law and International Law (Heidelberg), the Max Planck Institute for Meteorology (Hamburg) and the University of Hamburg. The School's research is focused on the legal, economic, and geophysical aspects of the use, protection, and organization of the oceans. Its researchers work in the fields of law, economics, and natural sciences. The School provides extensive research capacities as well as its own teaching curriculum. Currently, the School has 15 Directors who determine the general work of the School, act as supervisors for dissertations, elect applicants for the School's PhD-grants, and are the editors of this book series:

Prof. Dr. Dr. h.c. Jürgen Basedow is Director of the Max Planck Institute for Foreign Private Law and Private International Law; *Prof. Dr. Peter Ehlers* is the Director of the German Federal Maritime and Hydrographic Agency; *Prof. Dr. Dr. h.c. Hartmut Graßl* is Director emeritus of the Max Planck Institute for Meteorology; *Prof. Dr. Lars Kaleschke* is Junior Professor at the Institute of Oceanography of the University of Hamburg; *Prof. Dr. Hans-Joachim Koch* is Managing Director of the Seminar of Environmental Law at the University of Hamburg; *Prof. Dr. Rainer Lagoni* is Director emeritus of the Institute of Maritime Law and the Law of the Sea at the University of Hamburg; *PD Dr. Gerhard Lammel* is Senior Scientist at the Max Planck Institute for Meteorology; *Prof. Dr. Ulrich Magnus* is Managing Director of the Seminar of Foreign Law and Private International Law at the University of Hamburg; *Prof. Dr. Peter Mankowski* is Director of the Seminar of Foreign and Private International Law at the University of Hamburg; *Prof. Dr. Marian Paschke* is Managing Director of the Institute of Maritime Law and the Law of the Sea at the University of Hamburg; *PD Dr. Thomas Pohlmann* is Senior Scientist at the Centre for Marine and Climate Research and Member of the Institute of Oceanography at the University of Hamburg; *Dr. Uwe Schneider* is Assistant Professor at the Research Unit Sustainability and Global Change of the University of Hamburg; *Prof. Dr. Jürgen Sündermann* is Director emeritus of the Centre for Marine and Climate Research at the University of Hamburg; *Prof. Dr. Rüdiger Wolfrum* is Director at the Max Planck Institute for Comparative Foreign

Public Law and International Law and a judge at the International Tribunal for the Law of the Sea; *Prof. Dr. Wilfried Zahel* is Professor emeritus at the Centre for Marine and Climate Research of the University of Hamburg.

At present, *Prof. Dr. Dr. h.c. Jürgen Basedow* and *Prof. Dr. Ulrich Magnus* serve as speakers of the International Max Planck Research School for Maritime Affairs at the University of Hamburg.

Printing: Krips bv, Meppel, The Netherlands
Binding: Stürtz, Würzburg, Germany